AF349267

76

Fortschritte der Chemie
organischer Naturstoffe

Progress in the
Chemistry of Organic
Natural Products

Founded by
L. Zechmeister

Edited by
W. Herz, H. Falk,
G. W. Kirby, R. E. Moore,
and Ch. Tamm

Authors:
D. R. Adams, M. Brochwicz-Lewinski,
A. R. Butler

Springer-Verlag Wien GmbH

Prof. W. Herz, Department of Chemistry,
The Florida State University, Tallahassee, Florida, U.S.A.

Prof. Dr. H. Falk, Institut für Chemie,
Johannes-Kepler-Universität, Linz, Austria

Prof. G. W. Kirby, Chemistry Department,
The University of Glasgow, Glasgow, Scotland

Prof. R. E. Moore, Department of Chemistry,
University of Hawaii at Manoa, Honolulu, Hawaii, U.S.A.

Prof. Dr. Ch. Tamm, Institut für Organische Chemie der Universität Basel,
Basel, Switzerland

Library of Congress Catalog Card Number AC 39-1015

Typesetting: Thomson Press (India) Ltd., New Delhi

Printed on acid-free and chlorine-free bleached paper

SPIN: 10684589

With 57 partly coloured Figures

ISSN 0071-7886
ISBN 978-3-211-83165-6   ISBN 978-3-7091-6351-1 (eBook)
DOI 10.1007/978-3-7091-6351-1

Contents

List of Contributors

ADAMS, Dr. D. R., Department of Chemistry, Heriot Watt University, Edinburgh EH14 4AS, Scotland

BROCHWICZ-LEWINSKI, Dr. M., Department of Radiology, The Royal Infirmary, Edinburgh EH3 9YW, Scotland

BUTLER, Dr. A. R., School of Chemistry, University of St. Andrews, Purdie Building, St. Andrews, Fife KY16 9ST, Scotland

Nitric Oxide: Physiological Roles, Biosynthesis and Medical Uses

D. R. Adams[1], M. Brochwicz-Lewinski[2], and A. R. Butler[3,*]
[1]Department of Chemistry, Heriot Watt University, Edinburgh, Scotland
[2]Department of Radiology, The Royal Infirmary, Edinburgh, Scotland
[3]School of Chemistry, University of St. Andrews, Fife, Scotland

Contents

List of Abbreviations and Acronyms

AC	Acetylcholine
ARDS	Adult Respiratory Distress Syndrome
Arg	L-Arginine
CaM	Calmodulin
CAT	Cationic Amino Acid Transporter
cDNA	Complementary Deoxyribonucleic Acid
cGMP	Cyclic Guanosine Monophosphate
Cit	L-Citrulline
CNS	Central Nervous System
CPR	Cytochrome P450 Reductase
DAHP	2,4-Diamino-6-hydroxypyrimidine
DHFR	Dihydrofolate Reductase
EDRF	Endothelium-Derived Relaxing Factor
EDTA	Ethylenediamine Tetraacetic Acid
eNOS	Endothelial Nitric Oxide Synthase
EPR	Electron Paramagnetic Resonance
FAD	Flavine Adenine Dinucleotide
FMN	Flavine Mononucleotide
GSNO	S-Nitrosoglutathione
GTN	Glyceryl Trinitrate
GTP	Guanosine Triphosphate
GUK	Guanylate Kinase Homologous domain
HAP	Huntingtin-Associated Protein
H_2B	Dihydrobiopterin
H_4B	Tetrahydrobiopterin
HD	Huntington Disease
iNOS	Inducible Nitric Oxide Synthase
IL-1	Interleukin-1
IL-2	Interleukin-2
ITU	Intensive Care Unit
L-NAME	N-Nitro-L-arginine Methyl Ester
L-NMMA	N-Monomethyl-L-arginine
LPS	Lipopolysaccharide
mCPBA	*Meta*-Chloroperoxybenzoic Acid
MPAP	Mean Pulmonary Artery Pressure
NADH	Nicotinamide Adenine Dinucleotide
NADPH	Nicotinamide Adenine Dinucleotide Phosphate
NALA	N^G-Allyl-L-arginine
NANC	Nonadrenergic noncholinergic
NCPA	N^G-Cyclopropyl-L-arginine
NHA	N^ω-Hydroxy-L-arginine

NHAP	N-(N-Hydroxyamidino)piperidine
NHMA	N^G-Hydroxy-N^G-methyl-L-arginine
NMA	N^G-Monomethyl-L-arginine
NMDA	N-Methyl-D-aspartate
NNA	N^ω-Nitro-L-arginine
nNOS	Neuronal Nitric Oxide Synthase
NOS	Nitric Oxide Synthase
PAH	Phenylalanine Hydroxylase
PH	Pleckstrin Homology domain
PIN	Protein Inhibitor of nNOS
PKC	Protein Kinase C
PNS	Peripheral Nervous System
PPHN	Persistent Pulmonary Hypertension in the Newborn
PSD	Postsynaptic Density Protein
qH_2B	Quinonoid Dihydrobiopterin
ROS	Reactive Oxygen Species
SDS	Sodium Dodecyl Sulfate
SH3	src Homology 3 Domain
SNAP	S-Nitroso-N-acetylpenicillamine
SNC	S-Nitrosocysteine
SNP	Sodium Nitroprusside
SOD	Superoxide Dismutase

1. Introduction

The discovery that nitric oxide (NO) has an important role in animal physiology is one of the most exciting and interesting of recent years. The story began when it was discovered that NO is a messenger molecule in the cardiovascular system but, since then, many other roles have been assigned to this small and, apparently, insignificant molecule. Indeed, so much has been published on the subject that to prepare even a summary of the available material is a daunting task, made worse by the appearance of new material during the process of writing. Rather than attempting a comprehensive account this review, after a general introduction, will highlight three important aspects of the biological roles of NO: the enzyme nitric oxide synthase, S-nitrosthiols as carriers of NO, and the medical uses of NO. Other important topics involving NO will be mentioned only in passing and we apologise to workers in these areas for not giving their work the prominence it deserves but the length of this review must be kept within bounds. An effort has been made to provide key references to allow interested readers to pursue matters which capture their attention.

References, pp. 144–186

2. Discovery in the Vasculature

There are many oxides of nitrogen but only three are commonly encountered: nitrous oxide (N_2O), nitric oxide (NO) and nitrogen dioxide (NO_2). The first (N_2O) is used extensively in anaesthesia and is sometimes known as laughing gas. The third together with its dimer N_2O_4, is well known as the brown fumes given off by reaction of nitric acid with many substances and is present in city atmospheres as a pollutant. It is the second oxide (NO) which has been the subject of so much attention in biological, biochemical and medical circles during recent years. That such a well-known molecule should have been undetected in spite of extensive and intensive study of animal physiology is, at first, surprising but becomes more readily understood when the rather unusual properties of NO are taken into account. Although a radical (*i.e.* each molecule possesses an unpaired electron) it is generally unreactive. It does not dimerise, it reacts readily with ferrous iron and many radicals but with little else and in isolation is indefinitely stable. It is thermodynamically unstable with respect to the other oxides of nitrogen and this means that NO which has been stored for some time contains both N_2O and NO_2 because of the reaction:

$$3NO \longrightarrow N_2O + NO_2$$

Probably the best known reaction of NO in the gas phase is its ready oxidation by dioxygen to N_2O_4.

$$2NO + O_2 \longrightarrow 2NO_2 \longrightarrow N_2O_4$$

Subsequent hydrolysis of N_2O_4 gives equimolar amounts of nitrite and nitrate. In solution, on the other hand, the NO_2 formed on oxidation immediately reacts with NO to produce N_2O_3 which undergoes hydrolysis to give only nitrite (*1, 2*).

$$2NO + O_2 \longrightarrow 2NO_2$$

$$NO + NO_2 \longrightarrow N_2O_3$$

$$N_2O_3 + H_2O \longrightarrow 2HNO_2$$

The kinetic investigations upon which these equations are based established that, at concentrations applying in a biological situation, the rate of conversion of NO to NO_2 is slow enough for it not to interfere with the biological activity of NO and the eventual concentration of nitrite is a good measure of the amount of NO which was present in a biological fluid. In a biological situation NO does not give rise directly to nitrate.

To understand the experimental evidence that led to the discovery of NO in the vasculature it is necessary to appreciate a little of the

mechanism of vascular smooth muscle (VSM) relaxation. The term
smooth is simply a description of the physical appearance of muscles in
artery walls and may be contrasted with striated muscle which occurs in
the skeletal system. Contraction of VSM leads to vessel constriction and
is one of the factors influencing blood distribution and blood pressure.
Sustained contraction, in the absence of compensation elsewhere in the
body, can lead to high blood pressure (hypertension). Work by Murad
et al. (*3*) established that smooth muscle relaxation (which is a positive
process rather than just the absence of contraction) requires activation of
the enzyme guanylate cyclase and is accompanied by the conversion of
guanosine triphosphate (GTP) into cyclic guanosine monophosphate
(cGMP). The process of relaxation can be triggered by a number of
substances occurring in the body, including acetylcholine (Ac) and
bradykinin, and it had been generally assumed that they act directly upon
muscle cells. However, studies by Furchgott and Zawadzki (*4*) showed
that this is not the case. In the absence of the endothelium (the layer of
cells which lines the lumen of the artery) none of these substances is
fully effective. They concluded that acetylcholine acts, not upon muscle
cells, but upon the endothelium which produces a further messenger
molecule. This molecule then diffuses into the surrounding muscle cells
and activates guanylate cyclase. This messenger molecule became
known as the 'endothelium derived relaxing factor' or EDRF and its
chemical identify was a matter of much study and speculation.

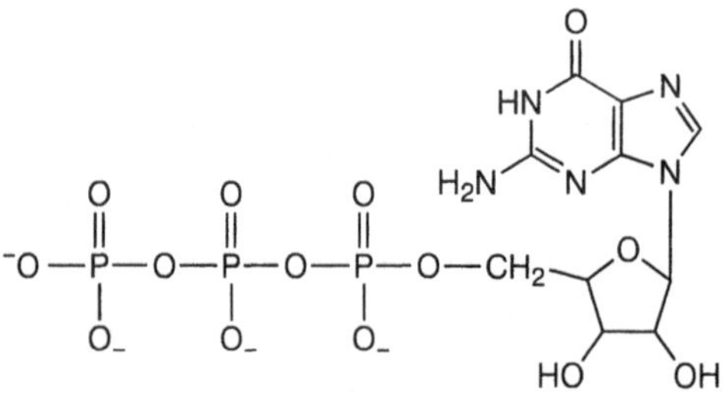

GTP

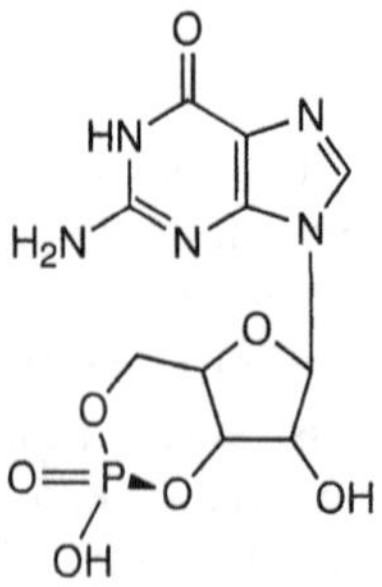

cGMP

Ac

It had been known for some time that guanylate cyclase can be activated *in vitro* by a number of compounds containing the NO group in some form or other, such as glyceryl trinitrate (GTN) and sodium nitroprusside (SNP) and, indeed, by NO itself. With hindsight the

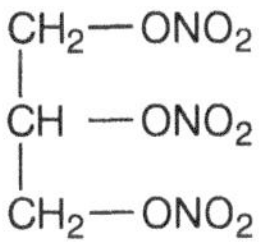

$$CH_2-ONO_2$$
$$|$$
$$CH\ -ONO_2 \qquad Na_2[Fe(CN)_5NO]$$
$$|$$
$$CH_2-ONO_2 \qquad\qquad SNP$$

GTN

chemical identity of EDRF should have been relatively easy to deduce but the idea of endogenous production of NO appeared, at the time, so unlikely. This was partly because its reactivity as a radical was overstated. However, a number of workers, including FURCHGOTT, suggested that NO was the endogenous EDRF but experimental proof was difficult to obtain. In 1987 two separate groups published the much sought-after evidence that EDRF was, indeed, NO. The British group, then working *a* the Wellcome Research Laboratories, was led by MONCADA (5) while the American group was led by IGNARRO (6). Both groups showed, by means of a bioassay, that certain properties of EDRF were identical to those of NO. It was quickly established that the precursor of NO is L-arginine, Arg, (7, 8) and this information led to the

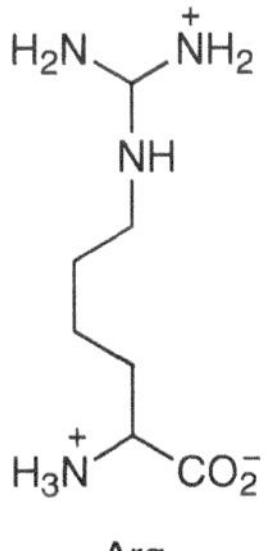

Arg

most direct evidence for endogenous NO production. Cultured endothelial cells were fed with L-arginine labelled with ^{15}N at the terminal position, perfused with a biological buffer and the perfusate purged with helium which was passed into a mass spectrometer. After stimulation of the endothelial cells with bradykinin the helium was found to contain NO labelled with ^{15}N (9). The early reports led to a flurry of activity on the L-arginine-NO pathway. The enzyme, or family of enzymes, responsible for NO production was named NO synthase and is considered in detail later in this review (Sections 8, 9, 10 and 12). A somewhat simplified view of what occurs when an endothelial cell is stimulated by arrival of an endothelium-dependent vasodilator like acetylcholine is shown in Fig. 1.

The way in which NO activates guanylate cyclase is now fairly well understood. The enzyme is found in most cells and throughout the animal kingdom. It exists in two forms: a soluble enzyme inside the cell and in a membrane-associated (particulate) form. The relative amounts of each within a cell vary with the cell type and its physiological state. Normally purified soluble guanylate cyclase can be activated *in vitro* by NO-donating compounds but if the purification is sufficient to remove the heme component of the enzyme, activation is markedly reduced (*10*). It can be restored by addition of hemeatin in the presence of a reducing agent. NO binds strongly to the iron of the heme group causing it to undergo a three-dimensional change (rather like the binding of dioxygen to the iron of hemoglobin) that increases the production of cGMP from GTP (see Fig. 2). Accumulation of cGMP in muscle cells leads, eventually, to relaxation (*11*).

Had NO been only a messenger molecule in the cardiovascular system its discovery would have been a major event in our understanding of biological chemistry. Subsequent events showed that this event was only the beginning of a set of discoveries which would revolutionise some aspects of biological chemistry.

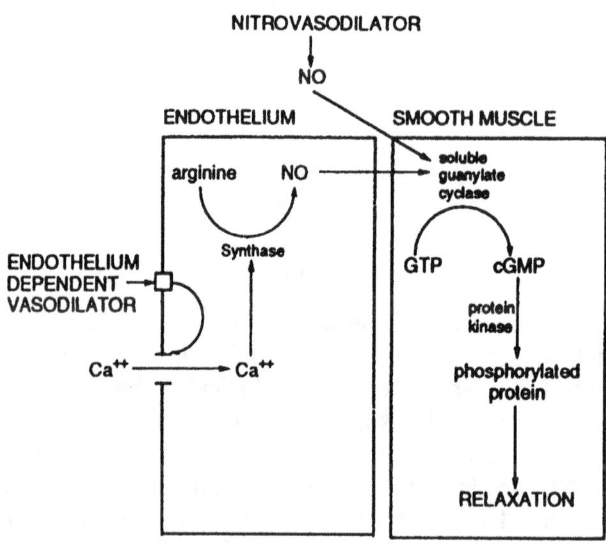

Fig. 1. NO production in endothelial cells

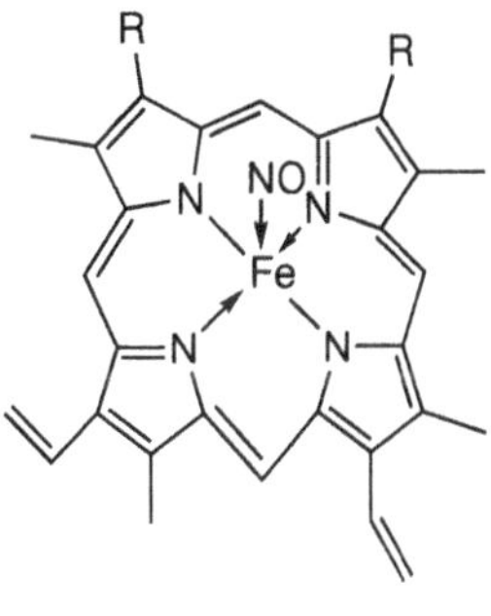

Fig. 2. Activation of guanylate cyclase

3. Platelet Aggregation

NO is involved in another important aspect of the blood supply. In blood there are numerous colourless cell fragments containing granules, known as platelets. They are much smaller than red blood cells. When a blood vessel is damaged, excessive bleeding is prevented by platelet aggregation to form a plug which adheres to the wall of the blood vessel. Further aggregation may lead to formation of a thrombus which can result in blockage of the blood vessel, particularly if it is coated with atherosclerotic plaque (a coating containing quantities of cholesterol on the inside of a blood vessel). Concurrently with the discovery of the role of NO in effecting vascular muscle relaxation came the discovery that NO inhibits both platelet aggregation and adhesion (*12*). Prostacyclin and NO act synergistically to inhibit platelet aggregation and to disaggregate platelets but there is no parallel synergism in platelet adhesion (*13*). The role of NO in this area seems to be as part of a feedback mechanism to counteract the effect of substances in the blood, produced after injury, which bring about aggregation and adhesion. The NO utilised by platelets is derived from endothelial cells with which the platelets come in contact, but there is also an enzyme system in platelets which uses arginine to produce NO. The transfer of NO from endothelial cells to platelets is something of a mystery as platelets are surrounded by cells containing hemoglobin and hemoglobin is one of the best known scavengers of NO. The scavenging of NO by hemoglobin is further considered in Section 6 in connection with *S*-nitrosothiols.

The arginine-NO pathway in human platelets has been extensively characterised (*14*) and it is known that a number of NO-donor compounds will inhibit platelet aggregation (*15, 16*). There appears to be the need for copper(I) ions in the antiplatelet action of *S*-nitrosoglutathione and it has been suggested (*17*) that this is due to the necessity for a copper(I)-containing enzyme. However, an alternative explanation is that copper(I) ions effect decomposition of S-nitrosothiols. The matter is discussed in detail in Section 6.

4. NO and the Immune System

The discovery of a role for NO in the vascular system was very important in itself and also had unexpected consequences. For example, it led an increased understanding of immune response. Immune response is the body's defence mechanism by which intrusive foreign matter, both living and non-living, is neutralised or destroyed. The non-specific immune response non-selectively protects against foreign substances or cells without having to recognise their specific identities. A key part of that response comes from a set of cells known as macrophages which are found in many tissues, their structure varying somewhat from location to location. For macrophages to respond to alien objects they have to be activated by substances known as cytokines. The role of macrophages is wide ranging and includes engulfing foreign matter (phagocytosis) and, if necessary, killing it by injection of cytotoxic substances. Macrophages can also kill invading microbes by contact without phagocytosis. It is also known that macrophages play a central role in non-specific death of tumour cells (*18*). It is the killing process which appears to involve NO.

It has been known for some time that adults or infants with diarrhoea excrete much more urinary nitrate than uninfected individuals (*19, 20*). More recent work (*21*) had shown that cultured macrophages from a mouse generate substantial amounts of both nitrite and nitrate after activation. It had also been shown that the cytotoxic activity of macrophages against tumour target cells depends upon the presence of L-arginine and is accompanied by formation of citrulline and nitrite (*22*). Once the production of NO from L-arginine in endothelial cells had been demonstrated it seemed probable that a similar process occurs in activated macrophages and that both nitrite and nitrate come from a common precursor, NO. This was confirmed by the work of three groups (*23–25*). However, the formation of nitrate from NO is something of a problem. In the gaseous phase in the presence of oxygen NO is rapidly and completely oxidised to NO_2, which hydrolyses to give both nitrite and nitrate:

$$2NO_2 + H_2O \longrightarrow HNO_3 + HNO_2$$

However, in solution in oxygenated water the reaction is different and only nitrite is produced (*1*):

$$4NO + O_2 + 2H_2O \longrightarrow 4HNO_2$$

No nitrate is produced in this reaction but nitrite is readily oxidised to nitrate in blood and this may explain its presence in urine. It now seems certain that, within macrophages, there is an enzyme system which is similar to that occurring in endothelial cells. It is unexpected that Nature should have chosen the same chemical system to act as a benign messenger in endothelial cells and as a cytotoxic agent in the immune system. However, the isoenzymes responsible for NO production are different in the two situations: that in endothelial cells is constitutive and that in macrophages is inducible. This difference is dealt with in detail in Section 8. It means that with macrophages NO is produced only when required while the vascular system requires continuous production of NO for basal processes and it is the quantity which is changed on demand. Also, much larger quantities of NO are produced by macrophages than by endothelial cells. There is, however, one diseased condition in which the two roles of NO cause serious difficulties. In septic or endotoxic shock a patient has a massive infection resulting in a great deal of activity of the immune system. The large quantities of NO produced have the serious side effect of substantially lowering blood pressure leading to hypotension (*26*) and the situation is frequently life-threatening. A preliminary study has shown (*27*) that NO synthase inhibitors can alleviate the symptoms of septic shock (see Section 11.7).

Why is NO a cytotoxic agent? Initially it was thought that because of its radical nature it was reactive enough to destroy or damage key cellular structures such as membranes but NO is, under most circumstances, a remarkably unreactive radical (*28*) and so this cannot be the correct explanation. Reaction of NO with dioxygen gives higher oxides of nitrogen which are nitrosating agents and can damage DNA (*29*) but it is not clear that this is part of normal immune response. However, reaction with another oxygen species may be part of normal immune resonse. Along with NO, the immune system produces superoxide (O_2^-) and, until the discovery of NO in mammalian physiology, it was thought that the cytotoxic species produced by macrophages were mostly reactive oxygen intermediates, of which superoxide is one. However, superoxide and nitric oxide undergo a very rapid reaction to produce peroxynitrite (*30*):

$$NO + O_2^- \longrightarrow OONO^-$$

This species is a powerful oxidant which may well be able to kill cells by destroying critical cellular components. Although attractive, this suggestion is not strongly supported by direct experimentation; the evidence is all indirect. Certainly the activity of NO as a vasodilator is increased by addition of the enzyme superoxide dismutase (SOD) (*31*), which is consistent with a lowering of steady state concentration of superoxide and, if formation of peroxynitrite is significant, consequent increased physiological lifetime of NO. Also it is known from *in vitro* experiments that peroxynitrite can effect nitration of tyrosine (*32*) and 3-nitrotyrosine has been detected in biological fluids where formation of peroxynitrite is likely (*33*). Peroxynitrite is a potent oxidising agent and reacts with a number of biologically important functional groups such as thiols and can initiate lipid peroxidation (*34*). As a generally destructive species peroxynitrite could be the basis of the cytotoxicity of NO. However, the lack of direct experimental evidence makes this conclusion far from certain. For peroxynitrite to act as a cytotoxic agent NO and superoxide must be produced simultaneously and in close proximity. If both species are produced by the same cell then that cell would be destroyed first unless it had a special protective mechanism. This matter is discussed perceptively by Fukuto and Ignarro in a recent review (*35*).

The reaction of peroxynitrite with tyrosine *in vitro*, and admittedly at rather high pH, has been examined by CIDNP techniques (*36*). The extent of tyrosine nitration was found to be very small and suggests that this reaction is not the origin of peroxynitrite's cytotoxic action, although it may indicate the presence of that species. The data obtained are consistent with the following pathway for the reaction of peroxynitrite in the presence of tyrosine at high pH:

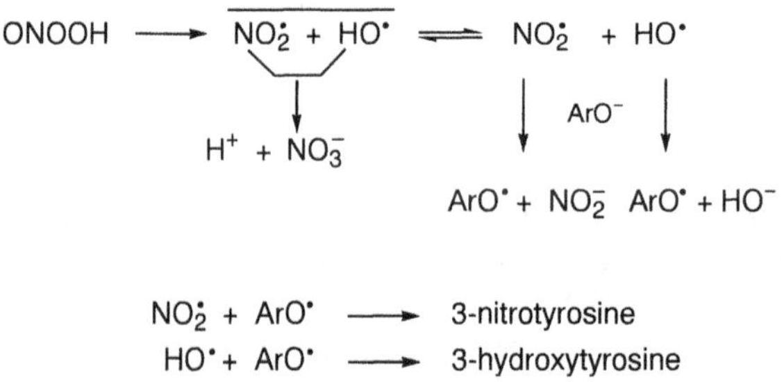

$$ONOOH \longrightarrow \overline{NO_2^{\bullet} + HO^{\bullet}} \rightleftharpoons NO_2^{\bullet} + HO^{\bullet}$$

$$\downarrow \qquad\qquad \downarrow\ ArO^- \ \downarrow$$

$$H^+ + NO_3^- \qquad ArO^{\bullet} + NO_2^- \quad ArO^{\bullet} + HO^-$$

$$NO_2^{\bullet} + ArO^{\bullet} \longrightarrow \text{3-nitrotyrosine}$$
$$HO^{\bullet} + ArO^{\bullet} \longrightarrow \text{3-hydroxytyrosine}$$

Fig. 3. Some *in vitro* reactions of NO with tyrosine

A horizontal bar indicates that the species are formed in a cage. The inverted and enhanced signals in the NMR spectrum indicate formation of some nitrate and some nitrite by radical processes and what little 3-nitrotyrosine is formed comes from reaction between NO_2, not peroxynitrite, and the tyrosinyl radical. There may also be a direct electron transfer from the tyrosine anion to peroxynitrite. One of the interesting properties of peroxynitrite is that it rapidly isomerises to the biologically benign species nitrate. If it is an important cytotoxic species one could see this as an advantage as any produced surplus to requirements will be rapidly neutralised by the isomerisation process. On the other hand, peroxynitrite formation *in vivo* might be seen as a way of regulating the biological activity of NO itself. To remove an excess of NO the body produces superoxide and the unwanted NO ends up as biologically inactive nitrate, with peroxynitrite as merely a transient intermediate. This view is discussed by FUKUTO and IGNARRO (*35*). Peroxynitrite formation cannot be seen as the complete answer to the problem of NO's cytotoxicity. It may be just one facet.

An important property of NO, which may be a major reason for its involvement in the immune system, is the readiness with which it reacts with iron, both heme and nonheme (*37, 38*). There is now direct evidence that generation of NO by macrophages produces a variety of iron-nitrosyl complexes (*39–41*). As there are a number of important iron centres in the enzymes and cofactors responsible for cellular metabolic processes, nitrosation of iron could bring about cessation of these processes leading to cell death. The biosynthesis of NO is known to modulate the activity of iron-dependent enzymes (*42*). Surprisingly, quite high concentrations of NO in aqueous solution have little effect on cultures of *Clostridium spirogenes* (*43*). However, species containing NO^+, such sodium nitroprusside $Na_2[Fe(CN)_5NO]$ (*44*) and Roussin's Black Salt $NH_4Fe_4S_3(NO)_7$ (*45*) are more effective antibacterial agents than NO itself. The significance of these observations may be that NO^+-containing species can readily nitrosate thiol groups, many of which occur in the active site of enzymes or are crucial in maintaining the 3D structure of proteins. *S*-Nitrosation of proteins may be a significant factor in understanding NO toxicity. Although NO cannot react directly with a thiol the necessary accompanying oxidative process could occur readily *in vivo* to give *S*-nitrosothiols. STAMLER *et al.* (*46*) have suggested that this process may serve as a signal transduction mechanism analogous to phosphorylation. The phrase "nitrosative stress" has been coined as a parallel to oxidative stress (*47*). FANG *et al.* (48) have studied the effect of *S*-nitrosoglutathione (a species which can release NO or transfer NO^+) on *Salmonella* mutants deficient in antioxidant defences. It has greatest activity for stationary cells by a mechanism involving NO^+ transfer rather than cell injury by NO. Further studies have shown that homo-cysteine, the only thiol intermediate in the *Salmonella* methionine bio-synthetic pathway, can function as an endogenous NO antagonist. These, and other studies, enforce the idea of signalling by S-nitrosation.

In summary, the exact role of NO in immune defence is still unclear. It cannot act as a generally destructive agent as it is too unreactive. Its cytotoxicity may result from conversion into peroxynitrite which is a much more reactive species, but which may lack the specificity expected of a good, endogenous cytotoxic agent. On the other hand, NO reacts very specifically with iron, a reaction which has much to commend it as a cytotoxic process. Equally NO, in an oxidative situation, may nitrosate important thiols thus disrupting certain key metabolic processes. The complete picture may be that all these processes, and possibly others, are involved in immune defence and the relative importance of any one may vary with the exact circumstances. Much more experimental work is required to resolve these matters.

5. NO and the Nervous System

Signalling between cells in the brain requires a neurotransmitter to carry the signal across synapses between cells. One such neurotransmitter is glutamate and exogenous glutamate is known to elicit large accumulations of cGMP in the cerebellum *in vivo* and *in vitro* (*49*), paralleling what occurs during relaxation of vascular smooth muscle. It is also known that arginine is the endogenous precursor for activation of rat forebrain guanylate cyclase and that activation can be inhibited by hemoglobin (*50*). In view of the similarities with vascular smooth muscle relaxation it seemed reasonable to look for NO. In 1988 Garthwaite *et al.* (*51*) reported that cultured brain cells produced an EDRF-like substance when stimulated by *N*-methyl-D-aspartate (NMDA), a synthetic amino acid which acts selectively at a subtype of glutamate receptor. At NMDA receptors, glutamate opens calcium ion channels, gatekeepers of neuronal transmission, thereby sending a strong excitatory impulse. The characteristics which linked the EDRF to the substance produced by NMDA stimulation include its short lifetime, its ability to raise cGMP levels, its sensitivity to nanomolar concentrations of hemoglobin and its ability to relax vascular smooth muscle. Once the EDRF had been identified as NO then it could be claimed that NO was produced, not only in vascular smooth muscle and during immune response, but also in the brain. After the observation by Garthwaite *et al.* the activity of an NO synthase in the brain was rapidly confirmed (*52*) and further details of neuronal NOS are described in Section 8. NO synthase from rat cerebellum was one of the first enzymes of this family to be purified and characterised (*53*).

As a neurotransmitter in the central nervous system NO is unique. It diffuses rapidly across biological membranes with no need for a carrier. Because of its short lifetime in a biological situation NO is not stored but immediately diffuses into a localised volume around its cell of origin. Its limited effect is due to the short lifetime. As calcium ions stimulate NO production, calcium is the link between extracellular stimulus (glutamate-NMDA receptor interaction) and intracellular generation of NO.

What is the role of NO in the brain? After NO has been produced by the postsynaptic neurone upon stimulation of a receptor it rapidly diffuses back to the presynaptic neurone and so can be seen as a feedback mechanism. It also diffuses towards other, neighbouring neurones thus establishing links with neurones other than those in the original signalling pathway (see Fig. 4). This behaviour has led to the suggestion that NO is involved in changes underlying learning and memory. Learning appears to involve increases or decreases in transmission across certain synapses after repetitive stimulation. Long term potentiation is part of a model for memory and is blocked by NO synthase inhibitors. What is lacking is a clear understanding of the exact role of cGMP in mediating the neuromodulatory action of NO. One of the most puzzling aspects of the role of NO in the brain is that as well as having a beneficial role (possibly in establishing memory) it is also neurotoxic during brain ischemia (restricted oxygen availability as in a stroke). The available data suggest that the impact of NO on the brain during ischemia is time dependent. Initially the effects of ischemia are countered by NO acting as a vasodilator and promoting cerebral circulation and microvascular flow. At the same time, glutamate-induced calcium overload in ischemic neurones leads to a persistent activation of neuronal NO synthase giving enhanced NO production leading to metabolic deterioration. Later, inducible NO synthase is expressed and the large amounts of NO produced lead to tissue damage (*54*). Further research here is of great importance as brain damage following ischemia is a cause of great human distress.

After the discovery of NO in the brain it was reasonable to look for the same substance in the peripheral nervous system. There is a family of

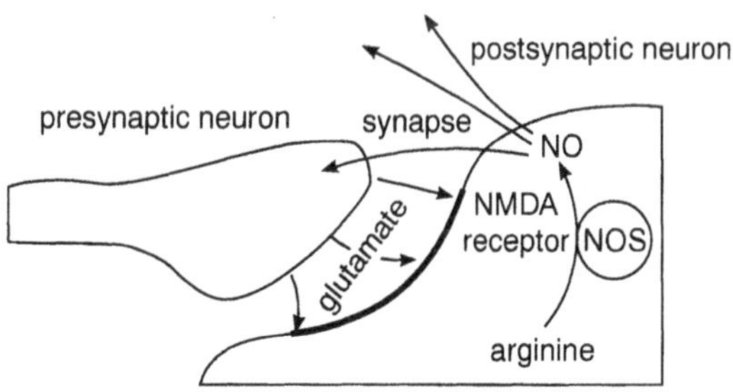

Fig. 4. Production and migration of NO in the brain

nerves known as nonadrenergic noncholinergic (NANC) nerves where the chemical nature of the neurotransmitter was, until recently, unknown. A role for NO was suggested by the observation that inhibitors of NO synthase block NANC neurotransmission (*55–57*). Also NO synthase has been found in nerve fibres that supply *inter alia*, the intestine, adrenal medulla and blood vessels (*58*). The evidence available so far suggests that NO plays an important part in nerves which control involuntary actions such as peristalsis and stomach size.

Much research had been carried out on the nature of the neurotransmitter in the NANC nerve supplying the bovine retractor penis (*59*). NO achieved some unexpected fame when it was announced that NANC neurotransmission caused relaxation of the human corpus cavernosum (responsible for penile erection) by mechanisms involving NO and cGMP (*60*). Research on the peripheral nervous system is still at an early stage but much evidence suggests that either NO is the NANC neurotransmitter or that the NANC neurotransmitter is a compound that stimulates muscle cells to generate NO.

6. *S*-Nitrosothiols

Any account of the physiological role of NO must include some comment on a class of compounds, hitherto of little consequence, but now of considerable importance because of their close relationship with NO: the *S*-nitrosothiols. Compounds of this class contain the grouping $-S-N=O$ and are the sulfur analogues of nitrites and, for this reason, are sometimes known as thionitrites. They are readily prepared by reaction of either acidified nitrous acid or N_2O_3 with a thiol (*61*):

$$RS^- + H_2NO_2^+ \longrightarrow RSNO + H_2O$$
$$RS^- + N_2O_3 \longrightarrow RSNO + NO_2^-$$

6.1. NO-Release

The reason for their close relationship with NO is that most *S*-nitrosothiols decompose readily with release of NO and formation of a disulfide (*62*):

$$2RSNO \longrightarrow RS-SR + 2NO$$

The manner in which NO release occurs is of great importance as it has been suggested that the EDRF is an *S*-nitrosothiol rather than NO, that

NO is stored in the body as *S*-nitrosothiols and that NO is transported around the body to where it is required as an *S*-nitrosothiol on hemoglobin. Each of these proposals will be discussed separately but first it is necessary to detail what is known of the chemistry of NO release from an *S*-nitrosothiol.

Initially it was thought that all *S*-nitrosothiols were unstable but that is not the case. *S*-Nitrosocysteine (SNC) can be prepared in solution but all attempts to isolate the compound result in decomposition. However, *S*-nitroso-*N*-acetylpenicillamine (SNAP) can be obtained (*63*) as green crystals which are indefinitely stable at room temperature (*64*). *S*-Nitrosoglutathione (GSNO) is also indefinitely stable (*65*). The release of NO from an *S*-nitrosothiol can be brought about in three ways. There

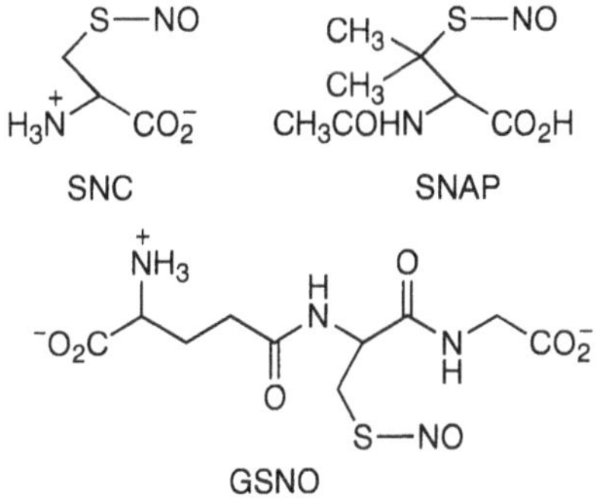

SNC

SNAP

GSNO

is a thermal reaction (*66, 68*) which, for many compounds of this class, is quite slow. On the other hand, photochemical release of NO (*67, 68*) is significant and should be considered whenever *S*-nitrosothiols are used in biological experimentation. However the process effecting most ready release of NO in solution is a metal ion-catalysed one (*69*). Copper is the most effective metal and there is enough in some samples of ordinary distilled water to bring about rapid NO release from an *S*-nitrosothiol. At first it was assumed that the more common Cu(II) ion was responsible for reaction but a more detailed study (*70*) showed conclusively that Cu(I) was the active catalyst. In most samples of an *S*-nitrosothiol there is enough thiol present as an impurity to convert all the adventitious copper present from Cu(II) to Cu(I):

$$2Cu^{2+} + 2RS^- \longrightarrow 2Cu^+ + RS-SR$$

It is the presence of this copper in distilled water which gave rise to the belief that many *S*-nitrosothiols are unstable in aqueous solution. If copper is removed by complexation with either EDTA or a chelating agent specific for Cu(I) ions such as neocuproine, then S-nitrosothiols are stable in solution over many hours as, in the absence of light, the only pathway for NO release is thermal decomposition. The consequence of this realisation is that it is quite reasonable to ascribe to S-nitrosothiols a biological role and to use them as NO-donor drugs. It is thought that Cu(I) acts as a catalyst by complexing with an S-nitrosothiol in the manner shown, thus weakening the S–NO bond. The matter has been considered by WILLIAMS (71) and other structures are possible. The presence of thiols has a complicated effect on the release of NO from a nitrosothiol as not only does the thiol reduce Cu(II) to Cu(I) so as to create the catalyst but the thiol may also complex Cu(II) and thus remove catalyst. One of the consequences may be induction periods in the release of NO (72).

$$CH_3\text{---}C(CH_3)(NHCOCH_3)\text{---}C(=O)\text{---}O^-\ \cdots\ Cu^+\ \cdots\ S\text{---}NO$$

Shortly after the first publications identifying the EDRF as NO appeared, an alternative identification as SNC was proposed (73). At the time there was considerable debate about this proposal and evidence against SNC and in favour of NO was published by MONCADA et al. (74). The problem with evaluating the evidence is that the decomposition of SNC to release NO, either as a thermal reaction or in a metal ion catalysed process, occurs so readily that distinguishing between the two is difficult. Although the debate about the exact chemical identity of the EDRF is not fully resolved the consensus at the time of writing is that what activates guanylate cyclase is NO but it is possible that NO is stored or transported as an S-nitrosothiol.

6.2. Endogenous S-Nitrosothiols

Some endogenous S-nitrosothiols are known. STAMLER et al. (46) have detected nitrosated human serum albumin in blood plasma and GSNO occurs in the lungs (75). These compounds could act as stores of NO for use in any physiological processes for which NO is required but there is, as yet, no conclusive experimental evidence for this. They also could represent waste products produced by NO not required for essential processes but their formation cannot occur by simple reaction of NO with a thiol (*76, 77*). Concurrent oxidation is required:

$$RS^- + NO \xrightarrow{[O]} RSNO$$

STAMLER *et al.* (*78*) have provided a most interesting suggestion for the use of an *S*-nitrosothiol in the transport of NO. In part it is an attempt to resolve one of the most puzzling features of the identification of NO as the EDRF. One of the best scavengers of NO is hemoglobin. NO bonds to the iron of heme more strongly than dioxygen and it is common in physiological experimentation to confirm the role of NO in the process under scrutiny by adding hemoglobin to the biological sample to show that attenuation occurs. The paradox is that *in vivo* NO is produced in close proximity to very large quantities of hemoglobin and some modelling by LANCASTER (*79*), based on Fick's second law of diffusion, has indicated that the amount of NO diffusing from endothelial cells to underlying muscle, rather than being scavenged by fast moving red blood cells in the opposite direction, is insufficient to exceed the K_m value for guanylate cyclase. The model is a very simple one and may not reflect fully the complexities of the *in vivo* situation but the general notion that much NO will be removed by blood from the site of production must be correct. STAMLER *et al.* (*78*) posit the idea that a thiol group of hemoglobin is nitrosated (to give Hb-SNO) in the lung when an allosteric change in hemoglobin, occurring on oxygenation, makes the thiol group more accessible. There is much NO activity in the lung (*80*) but the chemical process by which Hb-SNO is formed has not been established. It is also possible that, on oxygenation, any NO bound to iron as the result of NO scavenging is moved to form Hb-SNO. On reaching tissue where oxygen is released from oxyhemoglobin, the allosteric change results in NO loss from Hb-SNO as well. It is this NO which brings about vascular dilation rather than, or as well as, the NO produced locally. It is unlikely that locally produced NO, in spite of the scavenging paradox, plays no part in vascular dilation as administration of an NO synthase inhibitor (N^G-monomethyl-L-arginine) into the brachial artery, which supplies the forearm, brings about vasoconstriction (*81*). Also vessels do respond by dilation to local hypoxia (*82*) and to local shear stress (*83*).

6.3. *S*-Nitrosothiols as NO-Donor Drugs

In view of the simplicity of the chemical reaction leading to NO release from *S*-nitrosothiols it is not surprising that they are biologically active. The topic has been reviewed by STAMLER (*84*). They are potent vasodilators (*85, 86*) and differ in the effect in veins and arteries (87). There is general agreement that there is no obvious correlation between *in vitro* production of NO and vasodilator effect (*88, 89*). Indeed, KOWALUK and FUNG (*90*) have suggested that *in vivo* release of NO from RSNO is catalysed by an external vascular membrane. However, all previous studies of this type have ignored the dramatic effect of adventitious amounts of copper in the buffer used and must, therefore, be treated with some caution. *S*-Nitrosothiols are vasodilators in whole animals and are not subject to tolerance (*91, 92*). Another important property of *S*-nitrosothiols is that they prevent platelet adhesion and aggregation (*12, 13, 93*) (see Section 3). The effect of structure of an *S*-nitrosothiol upon antiplatelet activity has been examined (*94*). The antiplatelet activity of GSNO, from which spontaneous NO release is slow, has been examined (*15*) and there is evidence (*95*) to suggest that, for action, there must be cleavage of one peptide bond in GSNO by an enzyme acting like γ-glutamyltranspeptidase. GSNO appears to be more potent in antiplatelet activity than as a vasodilator and has been used clinically where this property is relevant. In a small chemical trial (*96*) GSNO has been used during percutaneous transluminal coronary angioplasty (balloon angioplasty), a procedure associated with platelet activation, acute vessel occlusion and chronic restenosis (cessation of blood flow). A single report (*97*) indicates that GSNO has beneficial antiplatelet activity in a pre-eclampsia variant known as HELLP syndrome associated with hemolysis, raised liver enzymes and low platelets levels (see Section 11.11).

In view of the *in vitro* sensitivity of *S*-nitrosothiols to Cu(I) ions for NO release, it is possible that *in vivo* NO release is triggered by the same catalyst. There is copper present in the body but most of it is complexed as Cu(II). However, work by DICKS and WILLIAMS (*98*) has shown that even complexed copper ions exercise catalytic activity. There are suffi-cient suitable reducing agents in animal tissue (cysteine, glutathione, ascorbic acid) to effect reduction of Cu(II) to Cu(I). The effect of copper ions on relaxations effected by nitrosothiols in the rat gastric fundus has been noted (*99*). GORDGE *et al.* (*100*) noted that addition of a complexing agent for Cu(I) reduced the antiplatelet activity of GSNO. In a subsequent paper (*17*) they suggest that there is a copper-containing enzyme responsible for NO release. Further work on this topic is awaited with interest. Neocuproine also attenuates the vasodilator action of GSNO and SNAP in *ex vivo* experiments (*101*). Any *in vivo* work involving the effect of Cu(I) upon NO release from GSNO is particularly difficult to interpret as even the *in vitro* situation is complex. There can be a long induction period before NO release commences and this induction period depends in a complex way upon the amount of oxygen present (*102*). At high concentrations of glutathione a copper-glutathione complex may form and there is also the possibility of a reaction leading to NO^- release (*103*):

$$GS^- + GSNO \longrightarrow GS{-}SG + NO^-$$

S-Nitrosothiols which do not complex Cu(I) for steric reasons are quite stable and the main reaction for NO release is thermal decomposition. This principle has been incorporated into the design of a vasodilator drug (RIG200). The acetylated glucose moiety provides the bulk

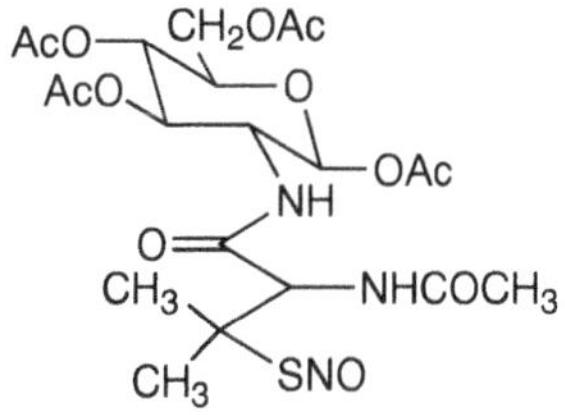

and a means of drug delivery. In *ex vivo* testing (*104*) RIG200 was found to be a slow release NO-donor compound which effects vasodilation over several hours. Similar compounds have been reported by WANG and his group (*105*) but the physiological effects of these compounds have not been reported. Roussin's Black Salt is another compound which lodges in vascular tissue and releases NO over several hours (*106*). It is possible that *S*-nitrosothiols will become important NO-donor drugs, particularly if they can be made tissue selective.

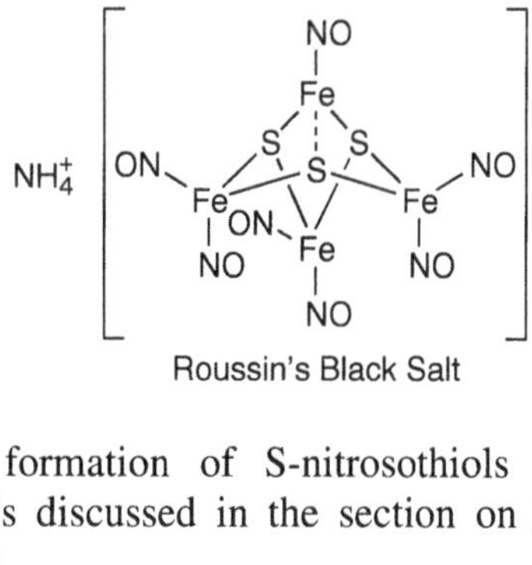

The possible formation of S-nitrosothiols under conditions of nitrosative stress is discussed in the section on NO and the immune system (Section 4).

7. NO Activity in the Mammalian Eye

To list the ways in which different organs utilise or are affected by NO would be one way of illustrating its diverse and multitudinous roles but such detail would be out of place in a review aimed specifically at chemists. Instead we have taken just one organ, the mammalian eye, and will describe ocular activities in which NO participates. Other organs use NO in as many ways and, by extension of what follows, it is not difficult to see the magnitude of NO's importance.

The supply of blood to the eye and the circulation of blood within the eye is regulated in a similar way to that in other organs and NO acts as a vasodilator here as elsewhere (*107*). However, as far as the eye is concerned, this function does not constitute the whole story.

Improper regulation of intraocular pressure has significant patholo-gical consequences. Although sufficient pressure is necessary to keep the eye spherical and to maintain optical clarity, excessive pressure may be a major risk factor for retinal damage. The optimal pressure is attained by a balance between secretion of aqueous humour by the ciliary processes and outflow of aqueous humour through the trabecular meshwork and eventually to scleral veins. It has been observed that small doses of nitrovasodilators such as GTN applied topically to the eye lower the intraocular pressure (*108*) due to changes in resistance to aqueous humour outflow (*109*). An extensive system of NOS-containing cells in ciliary muscle and outflow pathway has been detected (*110*). Clearly NO plays an important role in the control of intraocular pressure (*111*). Marked calcium-independent NOS activity is found in ciliary processes suggesting that NO production is highly regulated in this tissue (*112*).

Induction of NO synthesis by cytokines was demonstrated first in macrophages and, more recently in other cells including smooth muscle cells, chondrocytes, renal mesengial cells and hepatocytes. That list can now be extended to include human retinal pigmented epithelial cells (*113*). This clearly indicates that NO production is part of the ocular immune system. There are also NOS-positive cells concentrated in the inner parts of the ciliary muscle and this may indicate a role for NO in lens accommodation, *i.e.* the change in shape of the lens required to bring an object into sharp focus (*114*). The hemodynamic and vascular permeability changes associated with endotoxin-induced uveitis are mediated in large part by increased production of NO (*115*).

NOS is widely distributed in the bovine eye (*116*) and its presence in the retina (*117*) suggests that NO plays a part in retinal function possibly by influencing visual image processing through broad modulatory actions across the retina (*118*). This aspect of NO activity parallels that of NO in the central nervous system.

8. The NO Biosynthetic Pathway

8.1. Introduction

NO is generated in living organisms by oxidation of the guanidine function of L-arginine (Arg), a reaction which involves two successive monooxygenations* of the substrate and proceeds *via* an isolable intermediate, N^{ω}-hydroxy-L-arginine (NHA), to afford L-citrulline (Cit) and NO, Fig. 5. The entire process is catalysed by a remarkable family of enzymes, the NO synthases. Each monooxygenation step is thought to involve reductive activation of O_2 at a heme centre in the enzyme, with the electrons required for this being derived from the universal biological reductant, nicotinamide adenine dinucleotide phosphate (NADPH). In this review we refer to the two partial reactions of the scheme as nitric oxide synthase (NOS) monooxygenations I and II respectively.

The generally accepted stoichiometry for the overall process, as shown, involves consumption of 1.5 mole equivalents of NADPH and 2

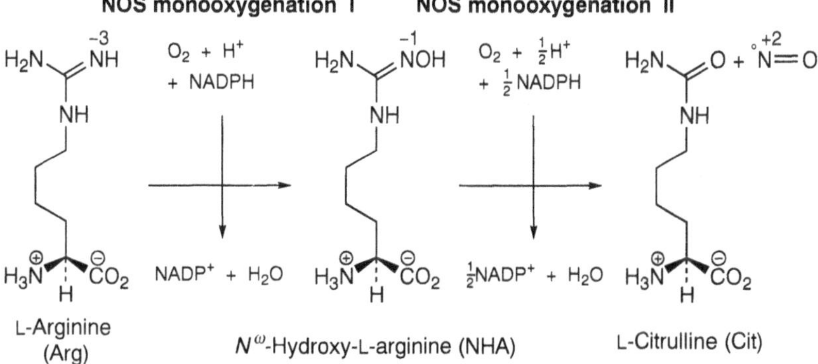

Fig. 5. The biosynthetic generation of NO from Arg consists of two successive monooxygenations, each consuming one molecule of O_2

---

*Enzymes that catalyse biological oxygenation reactions by direct incorporation of one or both oxygen atoms of dioxygen are called oxygenases. The classification is subdivided into two types: monooxygenases, which incorporate a single oxygen atom from dioxygen into the substrate, and dioxygenases, which incorporate both atoms.

$$2 \;[\text{H}_2\text{N–C(=NH)–NH–}] + 3\,\text{NADPH} + 3\text{H}^+ + 4\text{O}_2 \longrightarrow 2\;[\text{H}_2\text{N–C(=O)–NH–}] + 2\,\text{NO} + 3\,\text{NADP}^+ + 4\text{H}_2\text{O}$$

Fig. 6. Proposed overall stoichiometry for the oxidation of Arg to Cit and NO

equivalents of O_2, and constitutes a five-electron oxidation of the terminal guanidino N atom of Arg. This stoichiometry also invokes formation of two equivalents of water. Four electrons are formally required for O_2 reduction in each monooxygenation step and these electrons are partly derived from the substrate and partly from NADPH. Thus, in the first step, two of the required electrons derive from NADPH and the other two from the substrate: a formal two electron oxidation of $N(-3)_{\text{Arg}}$ to $N(-1)_{\text{NHA}}$. In the second monooxygenation, NADPH contributes a single electron and the substrate undergoes a formal three electron oxidation from $N(-1)_{\text{NHA}}$ to $N(+2)_{\text{NO}}$.

Table 1. *Milestones in the early characterisation of NOS and formulation of the Arg–NO pathway.*

1985	Immunostimulated macrophages were shown to produce NO_2^- and NO_3^-.
1987	Arg and Cit were identified, respectively, as substrate and co-product in the macrophage pathway producing NO_x^-.
	Macrophage-derived NO_x^- was found to issue from oxidation of one of the equivalent terminal guanidino N atoms of Arg.
	The intermediacy of NHA arising from an initial monooxygenation step in the Arg–NO_x^- pathway was postulated.
	Biosynthesis of NO in vasculature was announced and quickly linked to the macrophage Arg–NO_x^- pathway.
1988	NO was confirmed as an intermediate in the macrophage Arg–NO_x^- pathway and the nonspecific cytotoxic activity of macrophages was attributed to the production of NO.
	A common pathway for NO production in the vascular endothelium and in macrophages was demonstrated:
	(i) vascular NO, like macrophage-derived NO, was shown to derive from metabolism of Arg;
	(ii) the Arg derivative, N^G-monomethyl-L-arginine (NMA), known to inhibit production of both Cit and NO_x^- in macrophages was found to reduce the capacity of vascular endothelial cells for NO synthesis and for eliciting endothelium-dependent vasorelaxation.
	The existence of an NO-forming enzyme in endothelial cells was invoked. Synthesis of an NADPH-dependent cytosolic

Table 1. *(continued)*

	enzyme was also demonstrated upon induction of the macrophage Arg–NO_x^- pathway with immunostimulants. NO was shown to function as an intercellular messenger molecule in the brain. Neuronal production of NO was found to be linked to the activation of receptors by excitatory amino acids, triggering an influx of calcium ions which in turn were thought to activate the NO-producing enzyme.
1989	Endothelial NOS was provisionally established as a cytosolic, NADPH– and Ca^{2+}–dependent protein that exhibited high specificity for Arg, was inhibited by NMA, and was similar in many regards to the system responsible for NO production in macrophages. Preliminary characterisation of neuronal NOS also suggested a cytosolic, NADPH-dependent protein. The enzyme was inhibited by L-NMA but not D-NMA, and Cit formation required the presence of free calcium ions at physiological concentrations. The NOS requirement for a reduced biopterin cofactor was discovered.
1990	Neuronal NOS was purified to homogeneity. The Ca^{2+}/calmodulin-dependence of NOS was demonstrated.
1991	The NOS requirement for flavins was discovered.
1992	Heme was shown to be a NOS prosthetic group.
1990–92	NHA was conclusively shown to be an intermediate in the NOS-catalysed reaction pathway in the macrophage. The intermediacy of NHA was confirmed in further studies using NOS preparations from different sources. Observation of only trace amounts of NHA released during the enzyme-mediated reaction of Arg suggested that NHA was a tightly bound intermediate not normally released from NOS during NO biosynthesis. The N atom of NO was shown to be derived from the hydroxylated N of NHA, and NHA was shown to undergo oxidation exclusively at this N atom. The ureido O atom in Cit was found to derive from O_2 rather than water. Retention of the hydroxyl O atom of NHA in forming NO was shown. Analysis of the substrate-NADPH stoichiometry for formation of Cit suggested a requirement for 1.5 and 0.5 equivalents of NADPH for Cit formation from Arg and NHA respectively. Stoichiometric examination of NOS monooxygenation II revealed formation of an equivalent each of Cit and NO (measured as NO_2^- and NO_3^-) for each mole of NHA consumed.

Key steps in establishing the NO biosynthetic pathway are summarised below. The work is referenced in a review by GRIFFITHS and STUEHR (*119*).

With the basic features of the Arg–NO pathway in place, the major quest to elucidate the mechanistic detail of this important biological reaction began. According to the prevailing formulation the second NOS monooxygenation, like the first, is a mixed function oxidation in which the electrons required to reduce O_2 are derived partly from the substrate and partly from NADPH. However, unlike the first monooxygenation, which is thought to resemble conventional cytochrome P450 chemistry (*vide infra*), the second is without obvious biological precedent in that it involves a three-electron oxidation of the substrate and produces a radical product (NO). Indeed, a fundamental question concerns whether NO issues from the direct three-electron oxidation of NHA or whether a two-electron oxidation initially gives rise to a reduced nitrogen oxide species, nitroxyl (HNO) or its anion (NO^-), which subsequently suffers a further one-electron oxidation to produce NO itself. It is known, for example, that chemical oxidation of guanidoximes may produce either NO or HNO depending on the oxidising agent employed (*120, 121*).

The possible generation of NO^- rather than NO as the primary product of Arg turnover by NOS, prompted FUKUTO and others to investigate the pharmacological activity of nitroxyl (*122*). HNO, a weak acid of pK_a 4.7 (*123*), is expected to exist predominantly in the ionised form of NO^- at physiological pH. It was found that HNO generated, for example, by decomposition of sodium trioxodinitrate (Angeli's salt, $Na_2N_2O_3$) *in situ*, caused potent vasodilation of rabbit thoracic aorta by activation of guanylate cyclase. Although NO^- was found not to transform directly into NO in aqueous solution, weak oxidising agents such as Fe(III) hemoproteins were shown to mediate its one-electron oxidation and, therefore, the biological activity of NO^- containing solutions may have derived from the conversion of NO^- into NO.

Further work by FUKUTO, HOBBS and IGNARRO (*126*) demonstrated that NO^- is indeed readily oxidised to NO under physiological conditions by a variety of ubiquitous biological oxidants including O_2, superoxide dismutase (SOD)*, methemoglobin and flavins, although others (*125*)

---

*Superoxide dismutase is a copper-containing enzyme that catalyses the dismutation of superoxide (O_2^-) to H_2O_2 and O_2.

Angeli's salt ($Na_2N_2O_3$) decomposes to nitrite and the metastable species nitroxyl (HNO) in the pH range 3–9:

$$HN_2O_3^{\ominus} \xrightarrow{\text{pH 3-9}} NO_2^{\ominus} + [HNO]$$

Nitroxyl decomposes eventually to nitrous oxide through dimerisation to hyponitrous acid:

$$2HNO \xrightarrow[\text{of nitroxyl}]{\text{dimerisation}} \underset{HO}{\overset{OH}{N=N}} \xrightarrow[\text{hyponitrous acid}]{\overset{\text{decomposition}}{\text{of}}} N_2O + H_2O$$

Reduction of nitroxyl may produce hydroxylamine:

$$HNO \xrightarrow{[H]} H_2N-O^{\bullet} \xrightarrow{[H]} H_2N-OH$$

Fig. 7. Summary of nitroxyl chemistry [see (*124*) and (*125*) and references cited therein]

have been unable to detect NO release upon addition of H_2O_2 or O_2 to NO^- containing solutions. In particular, a 30-fold increase in the vasorelaxant potency of HNO in the presence of SOD compared to a 2-fold increase for the NO-donor compound SNAP strongly supported the hypothesis that the pharmacological activity of HNO issues from its ready conversion into NO. Murphy and Sies have also implicated SOD in the interconversion of NO^- and NO (*127*).

Such studies have fuelled a debate over whether EDRF is really NO itself, or whether another nitrogen oxide species such as NO^-, or a combination of species, can account for EDRF function. A recent study that compared the pharmacodynamic profiles of NO, NO^- and other candidate nitrogen oxide species, however, has concluded that there can be little doubt that NO is not only the species that activates guanylate cyclase but that NO is also the molecular form which carries the signal (*74*). Others have also observed that NO^- cannot by itself directly activate soluble guanylate cyclase but only after oxidation to NO (*125*), indeed that NO is the only nitrogen monoxide redox form capable of activating this enzyme (*128*). Observations that both the EDRF function and signal are mediated by NO do not, however, necessarily preclude the formation of NO^- as the immediate product of Arg turnover by NOS provided that rapid oxidation to NO occurs at or near its source of generation.

Significantly, recent work by Schmidt and co-workers will reignite the controversy over whether the proximal product of Arg turnover by NOS is NO or NO^- (*124*). These investigations demonstrated that in the absence of added SOD it was not possible to detect NO produced by purified NOS directly (using an NO–selective electrochemical sensor) *even under conditions when authentic NO introduced as an internal standard was completely recovered.* Moreover, two reduced nitrogen oxide species (N_2O and NH_2OH) were found to arise from Arg incubation with NOS in addition to NO_x^-. Formation of N_2O and NH_2OH, known markers for metastable nitroxyl (Fig. 7), strongly suggests generation of NO^- by turnover of Arg and cannot be accounted for by nonenzymatic degradation of NO.* NO^- formation would also be consistent with the requirement for SOD in order to observe an NO signal.

The results of the study by Schmidt and co-workers have yet further implications for the second NOS monooxygenation as currently formulated in the prevailing scheme depicted in Fig. 5. It was shown that peroxynitrite ($ONOO^-$), which is formed by the diffusion limited reaction of NO with superoxide, reacts rapidly with NADPH to form $NADP^+$ and NO_2^-. Thus superoxide formed in the assay medium[†] complicates the interpretation of results and suggests that the previously recorded NADPH:Cit stoichiometry of 1.5:1 has been substantially overestimated, with the corollary that additional reducing equivalents are required from a source other than NADPH for generation of NO. Unfortunately there is no direct and selective sensor for NO^- at present but these results clearly merit further investigation and may require reformulation of the NOS reaction.

8.2. Relation of Nitric Oxide Synthase to Cytochromes P450

Biological monooxygenation reactions are, of course, well known, and the heme-containing cytochrome P450 enzyme family in particular effect a wide range of such oxidations. These enzymes, originally discovered in liver microsomes in the late 1950s, are ubiquitous in nature–in mammals they are found in all tissues, being most abundant in the liver–and some 200 isoforms from various species have already been cloned and sequenced. Cytochromes P450 catalyse oxidation reactions on an extensive range of substrates (mostly hydrophobic), including the epoxidation of alkenes, hydroxylation of arenes and alkanes, oxidation and oxidative dealkylation of nitrogen and sulfur heteroatoms. In these reactions, one O atom from O_2 is inserted into the substrate and the other is reduced to water, Fig. 8. The main characteristic property shared by all

---

* See Addendum.

[†] Superoxide can be generated by autoxidation of one of the NOS cofactors, 2-amino-4-hydroxy-6-(L-*erythro*-1,2-dihydroxypropyl)-5,6,7,8-tetrahydropteridine (H_4B), added to the assay medium or by uncoupling of the NOS activation of O_2 from substrate oxidation.

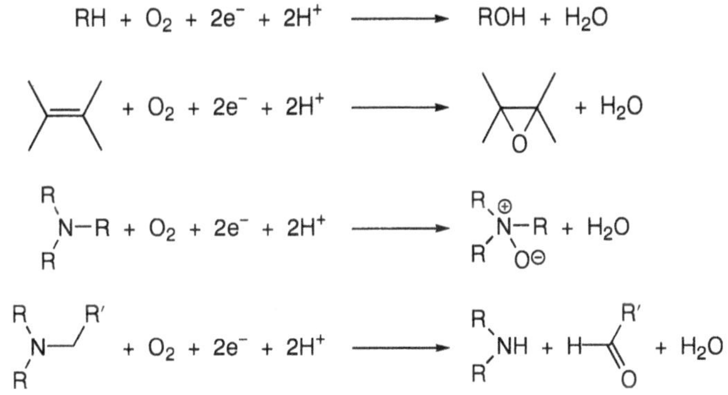

$$RH + O_2 + 2e^- + 2H^+ \longrightarrow ROH + H_2O$$

Fig. 8. Cytochromes P450, as part of multienzyme ensembles (monooxygenases), mediate biological oxidation reactions such as epoxidation of alkenes and arenes, *N*-oxidation, *N*-dealkylation and the chemically challenging hydroxylation of alkanes at unactivated positions

members of the cytochrome P450 superfamily is a distinctive absorption, the Soret peak, of their Fe(II)–CO complexes in the UV-visible spectrum around 450 nm from which they derive their name (*129, 130*).

Cytochrome* P450-catalysed substrate oxidation proceeds with stoichiometric consumption of a two-electron donor reductant, NADPH (or NADH), with electron transfer proteins facilitating the movement of electrons from reductant onto the heme moiety. Thus in their role as catalysts for biological monooxygenations, P450s are one component in a multienzymatic ensemble more accurately described as a cytochrome P450-dependent monooxygenase. Mitochondrial and most bacterial P450 systems comprise three components: a FAD-containing flavoprotein, an iron-sulfur protein and the P450. Microsomal P450-dependent monooxygenases, on the other hand, comprise a single flavoprotein, NADPH-cytochrome P450 reductase (CPR), in conjunction with the P450. CPR is composed of an N-terminal hydrophobic sequence, which serves as an anchor to the endoplasmic reticulum, followed by two subdomains that bind FAD and FMN flavin cofactors (*131*) (Fig. 9).

The majority of P450s so far examined comprise a single polypeptide chain of molecular mass around 50 kDa which binds an iron(III)protoporphyrin IX (heme) prosthetic group. However, in 1989 RUETTINGER

---

* The term *cytochrome* strictly refers to a hemoprotein which serves an electron and/or proton transport role by virtue of a reversible change in the valency of the iron. A hemoprotein possesses an iron porphyrin complex as the prosthetic group–most commonly iron(II)protoporphyrin IX. In the reduced Fe(II) state this complex is usually referred to as *heme* and in the oxidised Fe(III) state as *hemin*.

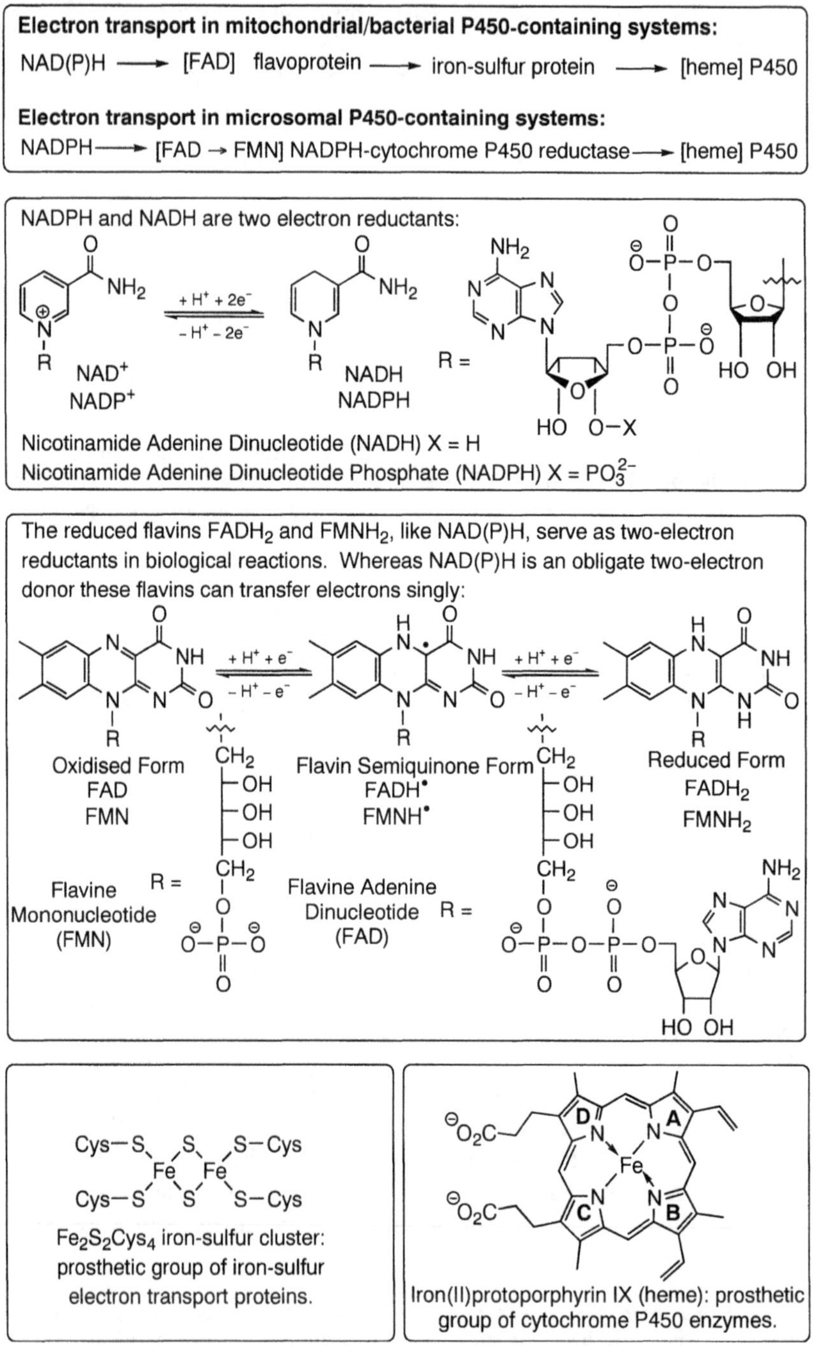

Fig. 9. Electron donors and components of electron transport pathways for P450-containing systems

et al. reported the amino acid sequence of a bacterial fatty acid P450 monooxygenase from *Bacillus megaterium*, P450$_{BM-3}$, with binding sites for both heme and flavin cofactors within a single, much longer polypeptide chain, and in which the reductase portion of the protein shows homology to CPR (*132–135*). More recently, independent disclosures from four laboratories in 1992 grouped the NO synthases, responsible for biosynthesis of NO, with the cytochrome P450 superfamily by identification of protoporphyrin IX as the prosthetic group of the enzymes (*136–139, 166*).

The isolation and subsequent molecular cloning of a NOS enzyme from rat cerebellum led to important insights into the structure and function of NOS (*53, 140*). In particular, the amino acid sequence deduced from the complementary DNA (cDNA) of the enzyme was found to contain consensus recognition sites for NADPH, FAD, FMN and the calcium-binding protein calmodulin (CaM). Indeed, the C-terminal section of the rat enzyme (*140, 141*), and all NOSs since studied, is similar to CPR, with 30–40% identity in the amino acid sequence and approximately 60% overall structural homology; consensus sequences for NADPH, FAD and FMN binding sites have been located (*140, 142, 143*). The flavoprotein character of the enzyme, which is homodimeric in structure and tightly binds one equivalent each of FAD and FMN per monomeric protein subunit, was confirmed by a number of groups (*138, 140, 144*–146). Several early reports (*147–154*) also described the dependence of NOS activity on levels of Ca^{2+} / CaM and another cofactor, namely 2-amino-4-hydroxy-6-(L-*erythro*-1,2-dihydroxypropyl)-5,6,7,8-tetrahydropteridine (H$_4$B). Since these early studies the cDNA sequences and predicted protein primary structures for NOS isoforms from several sources have been reported (*142, 155–165*).

Like cytochrome P450$_{BM-3}$, the NO synthases are distinguished from the majority of P450s by the coupling of heme oxygenase and P450 reductase domains within a single polypeptide. The identity and location of NOS cofactors within the amino acid sequence suggests that functionally NOS is a fused CPR-cytochrome P450 hybrid; for this reason NO synthases have been termed self-sufficient P450 enzymes, although the heme-binding oxygenase domain of the enzymes shows less homology to other P450s (*139, 167*). Indeed, NO synthases constitute the first example of a catalytically self-sufficient P450-type enzyme in mammals (*139*) and are the only mammalian enzymes apart from CPR known to contain both FAD and FMN.

NO synthases, then, are highly specialised cytochrome P450-like enzymes, unique in both complexity and their requirement for cofactors: in fact their catalytic function is achieved and regulated by a combination of no less than five bound cofactors and prosthetic groups: the flavins FAD and FMN, CaM, H_4B and heme (Figs. 9 and 10). Although these factors are widely associated with enzymes in nature, the NOS family alone appears to be uniquely distinguished by a simultaneous requirement for all five in order to function (*167–170*). In contrast to cytochromes P450, NO synthases are dimeric and possess two heme units per dimer. Each subunit of the dimer is characterised by a bidomain structure possessing both an oxygenase domain, containing the catalytic site at which reductive oxygen activation and substrate oxidation occurs, and a reductase domain, required for transmission of electrons from NADPH. The catalytic site in the oxygenase domain comprises the heme and pterin factors in close juxtaposition, while the flavins contribute to the reductase domain; the two domains are functionally linked through enzyme-bound CaM (*vide infra*). Evolution of this complex structure is thought to have arisen through convergent assembly of discrete polypeptide units with distinct ligand-binding and functional roles through duplication, rearrangement and fusion of pre-existing genes (*171*).

A calcium-binding, regulatory protein of 148 amino acids with binding sites for four Ca^{2+} ions. Calcium binding triggers a conformational change and so exposes a hydrophobic site which interacts with the protein (NOS) regulated by CaM.

Calmodulin
(CaM)

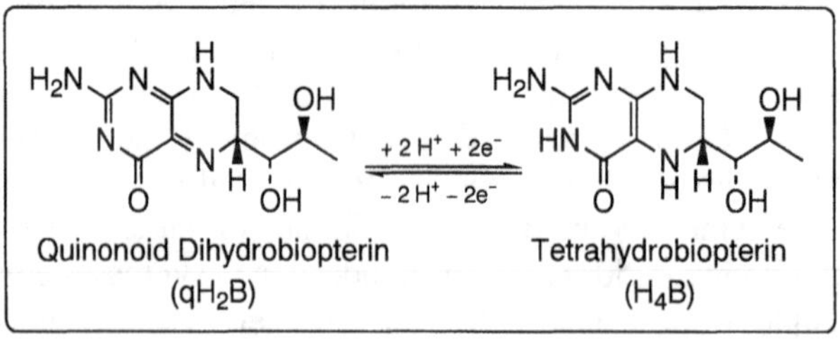

Quinonoid Dihydrobiopterin
(qH_2B)

$+ 2 H^+ + 2e^-$ / $- 2 H^+ - 2e^-$

Tetrahydrobiopterin
(H_4B)

Fig. 10. NO synthases constitute a uniquely complicated class of enzymes requiring a combination of five cofactors and prosthetic groups – heme, CaM, H_4B, FAD and FMN (see also Fig. 9) – for functional activity; reductive activation of dioxygen at the catalytic site, comprising heme and H_4B in close proximity, is achieved by oxidation of NADPH with electron transfer *via* FAD and FMN regulated by CaM

The NOS dependence on H_4B for maximal activity and stability further sets the NOS isozymes apart from other members of the P450 superfamily (*152, 172, 173*), for although H_4B is associated with iron-containing enzymes that catalyse hydroxylation reactions, the aromatic amino acid hydroxylases (*174, 175*), these enzymes are non-heme proteins. Indeed, the NOS requirement for H_4B and the lack of amino acid sequence homology of the NOS oxygenase domain with other P450s are considered by some authors to place the NO synthases out with the P450 super gene family (*176*). Nevertheless aspects of the NOS-catalysed reaction, Fig. 5, are thought to correspond to established cytochrome P450–mediated transformations, and for this reason a short perspective summarising cytochrome P450 chemistry is given in Section 9.1.

8.3. NO Synthase Isoforms

Soon after the discovery of NO biosynthesis it became apparent that the Arg–NO pathway is widely distributed in mammals and modulates a variety of biological consequences; the existence of a system of isoenzymes responsible for NO synthesis in different cell lines was, therefore, postulated (*52, 177*). Indeed, subsequent work has confirmed the existence of three NOS isoforms in mammals corresponding to the different tissue / cell types in which NO biosynthesis was first discovered *i.e.* an endothelial isoform (*178*), a neuronal isoform (*53, 149, 179*) and an inducible isoform associated with macrophages (*145, 146*). These isoforms are designated, respectively, eNOS, nNOS and iNOS*; the neuronal and endothelial isoforms are constitutively expressed in cells whereas iNOS is normally absent in resting cells, but is induced in response to bacterial products and immunoactive cytokines.

Mammalian NOS isoforms are the product of three distinct genes composed of 26 exons for iNOS (*163*) and eNOS (*180*) or 29 exons for nNOS (*181*). The enzymes are catalytically active as homodimers containing one equivalent each of heme (*136, 137, 139, 166*) FAD and FMN (*146, 182*) per protein subunit. The stoichiometry of H_4B per NOS monomer is prone to vary between enzyme preparations, but the macrophage isoform can contain full stoichiometric levels of this

---

* Other designations are also used in the literature: for the neuronal NOS–ncNOS, bNOS, NOS-1, NOS1, NOS-I; for the inducible isoform–mNOS, macNOS, NOS-2, NOS2, NOS-II; for the endothelial isoform–ecNOS, NOS-3, NOS3, NOS-III. Some favour the numerical nomenclature over the histological designations because the enzymes are not in fact restricted to the tissue types from which they were originally isolated.

cofactor (*152*). In contrast, nNOS reportedly contains one biopterin unit per NOS dimer (*151*). The three isoforms have similar biochemical properties and catalyse the same reaction, but are distinguished from one another by differences in their primary structure and molecular weight. The main difference between the constitutive and inducible isoforms, however, appears to lie in the method of regulation.

A major factor that regulates the activity of the NO synthases is the presence of Ca^{2+}/CaM. CaM binding to NOS functions as a switch which facilitates electron transfer from the reductase domain onto the oxygenase domain of the enzyme, permitting reductive activation of oxygen and consequent oxidation of Arg to NO and Cit (*183*). The constitutive isoforms are strictly Ca^{2+}/CaM-dependent and are thus physiologically activated by hormones or neurotransmitters that cause an influx and enhancement in the intracellular concentration of free Ca^{2+} ions. These isoforms produce low concentrations of NO as a signal for regulatory purposes, mediated for example by the interaction of NO with guanylate cyclase. Interestingly, whereas nNOS activity shows a striking dependence on Ca^{2+} ions, guanylate cyclase is inhibited by free Ca^{2+} ions in the physiological range. This feature was suggested by Moncada *et al.* (*52*) to constitute a control mechanism such that release of NO within a cell would affect guanylate cyclase in neighbouring target cells while minimising stimulation of the enzyme within the cell producing NO. It is likely that the regulation of constitutive NOS activity involves mechanisms other than Ca^{2+}/CaM-dependence alone. Indeed, recent work (*184*) has identified a 10-kilodalton protein inhibitor of nNOS (designated PIN) which inhibits nNOS by a direct physical interaction and shows a remarkable homology across diverse species of organisms; PIN binding to nNOS destabilises the enzyme's dimeric structure which is required for functional activity. All three isoforms are subject to phosphorylation (*149, 157, 182*, 185–193), and activity, at least for the endothelial isoform, has recently been found sensitive to regulation by enzymes that control levels of endogenous NOS inhibitors, N^{G}-monomethyl-L-arginine and dimethyl-L-arginine (*194, 195*).

The activity of the inducible NOS isoform, in contrast to eNOS and nNOS, is apparently independent of elevations in Ca^{2+} concentration, a characteristic that arises from its tight binding of CaM at normal physiological Ca^{2+} levels and even under nominally Ca^{2+}-free conditions (*196*). This isoform, once expressed, is effectively permanently switched on and provides a high-output source of NO at destructive levels as part of a non-specific cytotoxic/cytostatic immune response to invading pathogens and tumour cells. Regulation of NO production by iNOS occurs through a combination of mechanisms such as transcriptional control over the rate of iNOS gene expression (*197, 198*); control over the availability of Arg substrate through cationic amino acid transporter (CAT) proteins (*199–201*) and arginase activity (*202, 203*); and control over the biosynthesis of essential cofactors (H_4B) required for iNOS activity (*204*). Recent work also suggests that NO may regulate its own production by limiting both the expression and activity of the enzyme (Section 10.8).

Although early studies on NO synthases featured isolates from mammalian sources, the Arg–NO pathway is not in fact restricted to mammals; studies over recent years have shown that NO biosynthesis is a fundamental process found in diverse organisms throughout the phylogenetic kingdoms: mammals *(205)*, birds *(206, 207)*, fish *(208–212)*, invertebrates *(213–227)*, plants *(228–231)*, fungi *(228)* and bacteria *(232)*. Certain denitrifying bacteria, however, are also known to produce NO by a route other than the Arg–NO pathway; in these cases NO is formed by reduction of nitrite and further reduced to nitrous oxide by the mediation of nitrite and nitric oxide reductase enzymes *(233)*. Another bacterial pathway to nitrogen oxides, involving oxidative metabolism of ammonia, is also well documented *(234)*.

The activity of eNOS, which appears to be largely restricted in its tissue distribution to the vascular endothelium *(235)*, is linked with EDRF formation *(155–158)* and thus vasodilation and the maintenance of vascular tone. However, recent reports have also identified eNOS in kidney tubular epithelial cells *(236)*, cardiac myocytes *(237–239)*, platelets *(238)* and in CA1 neurons of the hippocampus *(240)*. The role of eNOS in kidney epithelium is unknown, but in CA1 neurons eNOS activity appears to be linked to the memory related phenomenon of long term potentiation *(241, 242)*. Deficiency of this enzyme is linked to hypertension and vascular disease *(243, 244)*.

The neuronal isoform of NOS, originally localised in a subset of central neurons *(140)* following the discovery of NO biosynthesis in the cerebellum *(31–53, 245)* has a far more extensive tissue distribution than its endothelial counterpart. Thus, nNOS has now been identified by immunological staining techniques in several other cell types including skeletal muscle tissue, peripheral NANC neurons, pancreatic islet cells, kidney macula densa cells and some epithelial cells *(141, 161, 246–248)*. The activity of nNOS is associated with signal transduction in central and peripheral neurons *(53, 249)*. In human skeletal muscle nNOS is actually expressed at higher levels than in the brain *(161)*; a number of functions have been ascribed to NO in muscle including a role in opposing the contractile force of muscle fibers *(148, 250, 251)*.

Neither is the inducible NOS isoform confined to the cells in which it was originally identified (macrophages). In fact this isoform may perhaps be expressed in any nucleated cell subjected to an appropriate stimulus delivered by inflammatory cytokines or LPS. Thus iNOS has been identified in macrophages (*145, 146, 252*), hepatocytes (*163*), endothelial (*253*) and smooth muscle (*254*) cells of vascular endothelium, chondrocytes (*162*), myocardium (*255*), microglia (the resident macrophages of the brain) (*256*) and astrocytes (*256, 257*). The primary role of iNOS appears to be the killing or suppression of tumour cells and intracellular parasites and pathogens including viruses (*197, 205, 258, 259*), but prolonged or excessive induction of NOS is associated with pathophysiological conditions. For example, iNOS induction in neurons may contribute to Alzheimer's disease pathogenesis (*260*). It has been suggested that the inducible isoform may effectively be expressed constitutively in certain tissues (*197, 261*), in which case iNOS is probably best understood as that NOS isoform which is independent of Ca^{2+} levels (*197*). A recent review by WOLF discusses histochemical localisation of NOS (*262*).

9. Mechanism of the Nitric Oxide Synthase-Catalysed Reaction

9.1. Mechanisms of Cytochrome P450-Mediated Oxidations

Much of the current thinking about the NOS-catalysed Arg–NO pathway relates the catalytic mechanism of NOS to that of cytochromes P450. In particular it has been suggested that NOS monooxygenation I corresponds to a typical P450-mediated hydroxylation reaction and that the processing of some NOS inhibitors by the enzyme may proceed by a mechanism analogous to P450-mediated *N*-dealkylations. Parallels have also been drawn between the mechanism of the P450 enzyme aromatase and NOS monooxygenation II. Lastly, a P450-type epoxidation of the substrate guanidine group leading to labile oxaziridine intermediates might be considered as an alternative mechanistic pathway for NOS monooxygenations I and II. It is, therefore, necessary to preface a discussion of the NOS catalytic mechanism with a synopsis of cytochrome P450-mediated oxidations.

9.1.1. P450-Mediated Hydroxylation

The generally accepted mechanism for cytochrome P450-mediated hydroxylation involves hydrogen abstraction from the substrate by a high-valent oxoiron intermediate to generate a carbon-centred radical, the oxoiron species itself having been formed by reductive activation of O_2 at the heme-iron centre of the enzyme (*130, 263*). Once formed, the substrate-derived radical captures hydroxyl from the iron centre in an "oxygen rebound" step that produces the product and regenerates the Fe(III)-heme group of the enzyme, Fig. 11 (*129, 264, 265*). The electrons required for the reductive activation of O_2 in this cycle are provided mainly by NADPH through an endoplasmic reticulum-organised electron transport chain involving proteins such as CPR.

A soluble, bacterial camphor-hydroxylating enzyme, P450$_{cam}$ from *Pseudomonas putida*, was the first member of the cytochrome P450 enzyme family to be crystallised and have its structure determined (*266, 267*).* In this enzyme the heme prosthetic group is bound between two protein helices by a combination of hydrophobic interactions, hydrogen bonding / coulombic interactions between the heme propionate side-chains and protein Arg / His residues, and by cysteine thiolate axial

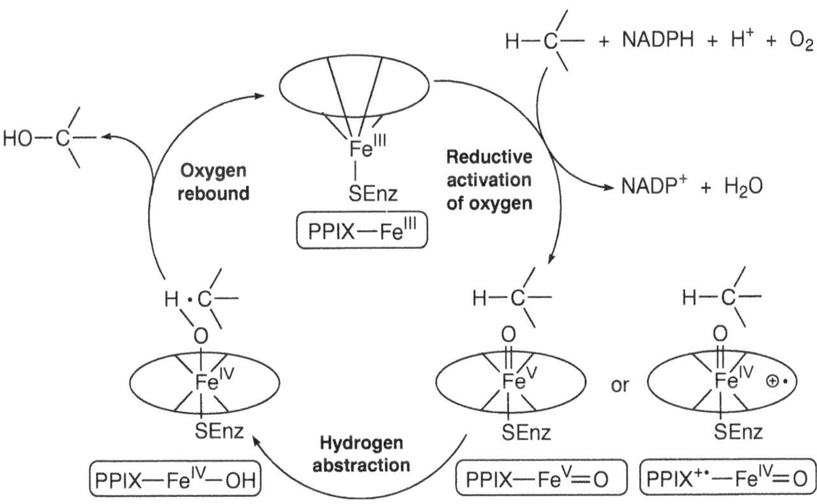

Fig. 11. Schematic representation of the cytochrome P450-catalysed hydroxylation reaction; the ellipsoid represents the porphyrin, SEnz represents the axial cysteine thiolate ligand donated by the protein

---

* The crystal structures of three other bacterial P450s have since been published: P450$_{TERP}$ (*268*), P450$_{eryF}$ (*270*) and P450$_{BM-3}$ (heme domain) (*269*).

ligation of the iron. The proximal thiolate ligand* may facilitate O–O bond cleavage leading to the high-valent oxoiron intermediate responsible for substrate oxidation. Camphor substrate is located approximately 4 Å from the plane of the porphyrin, directly adjacent to the oxygen binding site, and bound, partly by a hydrogen bond from its keto group to a tyrosine residue and partly by van der Waals interactions, in a cavity lined by hydrophobic residues. A threonyl hydroxy group in the dioxygen binding pocket may contribute to lysis of the O–O bond in forming the hypervalent oxoiron intermediate (271). The detailed electronic structure of this intermediate in P450-mediated reactions has been the subject of intense debate over the years; although originally formulated in 1966 as an oxoiron(V)protoporphyrin IX complex by ULLRICH and STAUDINGER (272), it is now widely accepted that the porphyrin ligand donates an electron to the iron forming an oxoiron(IV)porphyrin radical cation as the actual intermediate (273). Electron density may also be transferred from the axial thiolate ligand onto the heme iron, which accounts for the unusual red shift in the position of the Soret peak for cytochromes P450 from the normal position around 420 nm for non-thiolate ligated hemoprotein-Fe(II)-CO complexes (274).

In the case of $P450_{cam}$ the iron-sulfur redox protein that transfers the electrons required for reductive activation of O_2 onto the heme is putidaredoxin. Several basic amino acids on the proximal surface of $P450_{cam}$ are thought to facilitate binding of putidaredoxin. However, the heme prosthetic group is not directly accessible from the surface of the protein, the closest approach being approximately 8 Å from its proximal

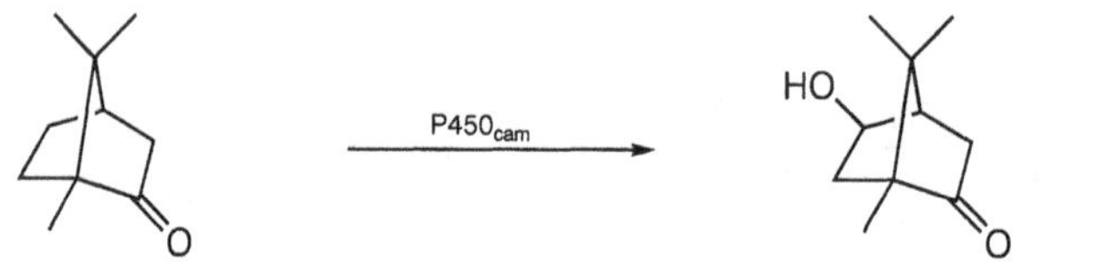

Fig. 12. The bacterial cytochrome P450, $P450_{cam}$, mediates stereoselective hydroxylation of D-camphor at the 5-exo position

---

* Iron(II) possesses a d^6 electronic configuration, thus in four-coordinate iron(II)protoporphyrin IX there are two vacant coordination sites. Additional ligands are constrained by the macrocycle to axial positions on either side of the porphyrin. In cytochrome P450 enzymes a cysteine thiolate ligand occupies one of these axial positions on the "proximal" side of the ring. Dioxygen coordinates in the remaining axial position, which is termed the "distal" site.

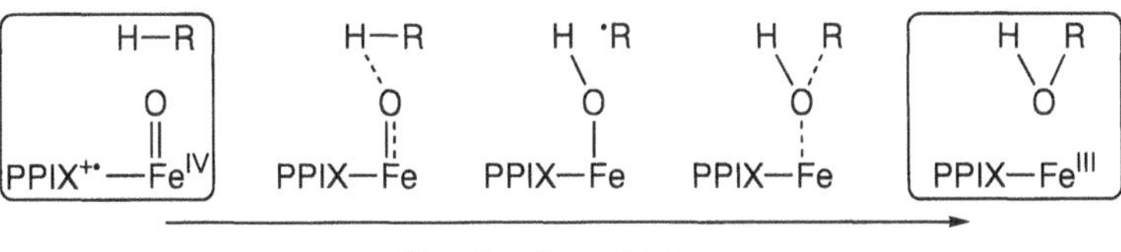

Fig. 13. Recent evidence (*275*) suggests that the carbon radical formed in the hydrogen abstraction-oxygen rebound hydroxylation of saturated hydrocarbons is so short-lived that it cannot be considered as a discrete intermediate in the reaction pathway

surface. Thus it is likely that one or more amino acid side chains conduct electrons from the redoxin onto the P450 heme (*130*).

Evidence accumulated over the last two decades from numerous experiments favours the two-step hydrogen abstraction-oxygen rebound mechanism for cytochrome P450-mediated hydroxylation at saturated carbon centres (*129*). However, recent studies on the mechanism of hydroxylation suggest that the intermediate radical generated by H abstraction from the substrate may be so short-lived in nature that, rather than existing as a true intermediate, it constitutes part of a reacting ensemble (*275*). Thus the H abstraction and oxygen rebound steps of the catalytic cycle essentially occur together as a non-synchronous concerted oxygen insertion process in which C–H bond cleavage leads C–O bond formation (Fig. 13). Such a mechanism requires "side on" approach of the C–H bond as opposed to a linear FeO...H–C array for a more conventional abstraction. Other recent work has also supported concerted hydroxylation mechanisms that proceed *via* a 5-coordinate carbon species (*276, 277*).

9.1.2. P450-Mediated N-Oxidation and N-Dealkylation

The reactive cytochrome P450 oxoiron intermediate also mediates *N*-oxidation of tertiary amines, although *N*-dealkylation competes in the case of amines that possess α-hydrogen atoms. Single electron transfer from the amine nitrogen atom to PPIX⁺˙–Fe(IV)=O with subsequent O capture by the resulting aminium radical cation accounts mechanistically for the *N*-oxidation process (*129, 264*), pathway A Fig. 14, and a similar mechanism might operate in the first NOS monooxygenation of Arg to Cit.

N-Dealkylation of tertiary amines may proceed *via* the same aminium radical cation invoked in *N*-oxidation (pathways B and C) or may alternatively involve direct hydroxylation at the α-carbon followed by collapse of the resulting carbinolamine (pathway D). These reactions

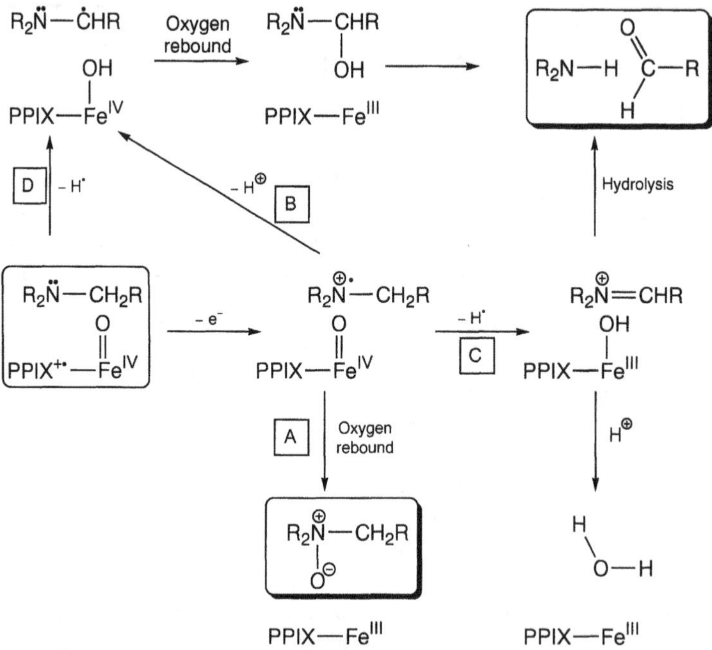

Fig. 14. Mechanisms for cytochrome P450-mediated *N*-oxidation (*Pathway A*) and *N*-dealkylation (*Pathways B-D*) of tertiary amines

are relevant to the processing of certain mechanism-based inhibitors by NOS, Section 9.4. Although pathways B and D are similar, the latter involves direct α-hydrogen abstraction from the substrate to generate a carbon-centred radical while the former gives rise to the same radical intermediate in a stepwise fashion. The stepwise process involves single electron oxidation followed by *C*-deprotonation of the aminium radical cation and then oxygen rebound. The debate over which pathway(s) are operative in P450-mediated *N*-dealkylations is summarised in Fig. 15; pathway B is generally favoured as the operative mechanism. In practice, however, it is frequently difficult to eliminate conclusively alternative mechanistic pathways and the situation is complicated by the fact that electron transfer from substrate to the perferryl heme intermediate is possible over reasonable distances (*278*). It has been estimated (*279, 280*), for example, that outer-sphere electron transfer, which does not require orbital overlap and the strict structural demands of an inner-sphere transfer mechanism, may occur over a distance of 5 Å or more in P450-mediated oxidations, although some (*281, 282*) regard the P450 redox potential required for it to act in such a manner to be infeasibly high.

- Aminium radical cations, which possess weakened α-C–H bonds, are relatively acidic in nature because their deprotonation leads to a resonance stabilised α-amino radical (pathway B). pK_a values of 9–15 have been estimated for aminium radical cations of tertiary amines (*283–285*).
- The relatively low ionisation potential of N and stability of the α-amino radical may conspire to favour *N*-dealkyation by pathway B (*285*).
- Evidence favouring pathway B and the role of PPIX–Fe(IV) = O as a base (*286*) which deprotonates the aminium radical cation has been reviewed by GUENGERICH and co–workers, who have suggested that this is the dominant pathway for oxidation of substrates with low $E_{1/2}$ values (< 1.5–2.0 V) (*287, 288*).
- Others have suggested that loss of a hydrogen atom to generate an immonium ion, a more stable species than the α-amino radical, is favoured for decomposition of non-biologically generated aminium radical cations (*283*). Thus pathway C, commencing with single electron transfer and proceeding with α-hydrogen abstraction from the aminium radical cation might be expected to operate.
- Studies of ^{18}O incorporation patterns in P450–catalysed dealkyations are consistent with a mechanism in which the aldehyde oxygen is not derived from water (pathways B and D) (*287*).
- Arguments favouring the operation of pathway D (*289–291*) and pathway B (*292*) have been presented. Generally pathway B is thought to be operative in P450-mediated *N*-dealkylations (*287, 292*).
- Direct detection of aminium radicals in NADPH-supported P450 reactions has not been reported, but GUENGERICH *et al.* were able to detect coloured aminium radical intermediates in P450-mediated demethylations supported by iodosylbenzene (PhIO).*
- Some recent publications have suggested that amine dealkylation can proceed *via* direct carbon oxidation through hydrogen atom abstraction (pathway D), in particular that P450-mediated transformations of sparteine, nicotine and 3-methylindole proceed by such a pathway (*278, 282, 293*). Thus it is probable that P450s can catalyse amine *N*-dealkylations by direct α-hydrogen atom abstraction when such a course is sterically or electronically favoured: clearly substrate orientation with respect to the perferryl porphyrin intermediate as well as the substrate's oxidation potential must be deciding factors in which mechanism operates in a given case.

Fig. 15. Leading arguments in the debate over the mechanism of P450-mediated
N–dealkylation of tertiary amines

The suggested *N*-oxidation and *N*-dealkylation mechanisms (pathways A and B) presuppose reaction of the substrate as the free base in order to proceed by initial electron transfer. A non-polar environment within the protein may serve to partition the amine from bulk solution in its non-protonated form which undergoes oxidation. Alternatively, if the

---

* Iodosylbenzene is a simultaneous O_2 and two-electron donor surrogate that is commonly employed in model porphyrin studies to generate the high-valent perferryl PPIX$^{+\cdot}$-Fe(IV)=O intermediate directly from PPIX-Fe(III) in the absence of O_2 and NADPH.

substrate accesses the P450 catalytic site as an ammonium ion, its deprotonation may be facilitated by the protein, liberating the free base required for electron transfer. An interesting case is P450 2D6 (debrisoquine 4-hydroxylase), a unique cytochrome P450 that typically effects oxidation of basic nitrogen-containing substrates, such as the antihypertensive guanidine-containing drug debrisoquine, at sites distant from the protonated nitrogen (*294–297*). MACDONALD and co-workers have proposed that *N*-dealkylation of the therapeutic agent L-deprenyl effected by this enzyme (Fig. 16) involves entry of the substrate into the catalytic site in its protonated form (*280*). The substrate is bound by a salt bridge to a protein carboxylate residue, but exists in equilibrium with its free base form, the latter suffering rapid electron transfer oxidation as the initiating step towards *N*-dealkylation. Two groups have recently identified a glutamate residue in the NOS heme distal pocket which

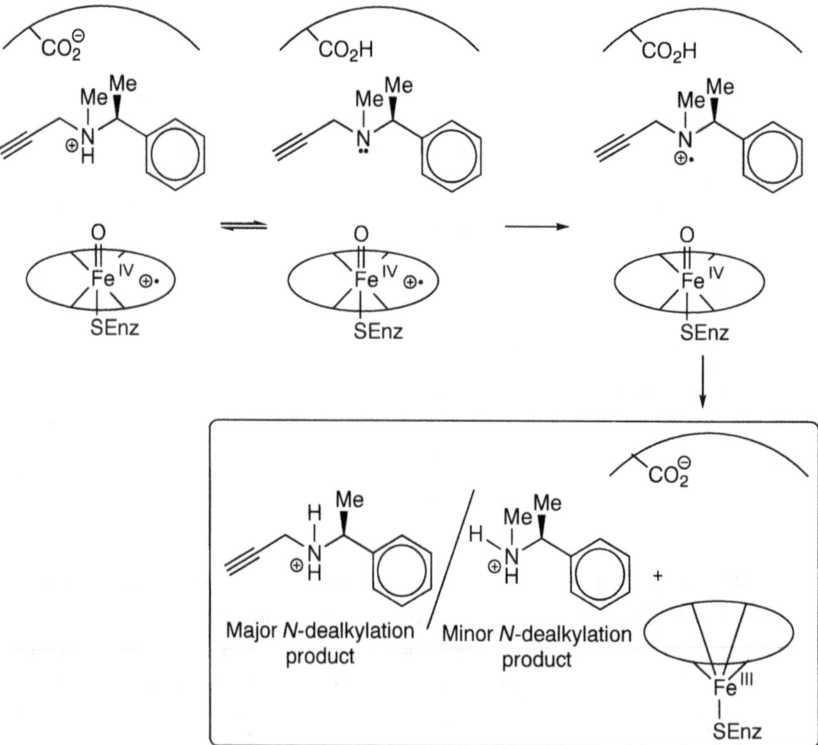

Fig. 16. *N*-Oxidation and *N*-dealkylation reactions effected by cytochromes P450, such as the *N*-dealkylation of L-deprenyl by P450 2D6 shown here, must operate on the substrate in its free base form if the mechanism commences with electron transfer

forms a salt bridge with the Arg guanidinium group (*298, 299*). However, a NOS monooxygenation I mechanism involving H atom abstraction rather than electron transfer has been assumed by one of the groups (*298*).

9.1.3. *P450-Mediated Epoxidation*

Carbon-carbon double bond epoxidation is thought to involve a similar high-valent oxoiron intermediate to that invoked in P450-mediated hydroxylations. By analogy to epoxidation, oxygenation of the imino double bond of the guanidine moiety to generate intermediate oxaziridine species may be considered as a possible mechanism in the NOS reaction (Section 10.9). It is generally accepted that olefin epoxidation commences with formation of a charge-transfer complex between the oxoiron intermediate and the substrate. Mechanistic steps thereafter have been the subject of intense debate. Observations of substrate rearrangements and heme alkylation have been taken as evidence of reaction pathways involving radical and/or cationic intermediates. In a recent review SONO *et al.* (*176*) have concluded that the substrate in the initially formed charge transfer complex suffers electron transfer to form a radical cation intermediate. The latter can proceed to epoxide product through either radical or cationic inter-mediates (Fig. 17). OSTOVÍC and associates (*273*), on the other hand, have dismissed the involvement of a radical cation intermediate *en route* to epoxides (*273*). In their interpretation of available data the most likely mechanism of epoxidation involves either a concerted insertion of the metal-bound oxygen into the alkene double bond or its electrophilic addition to the alkene to generate a carbocation intermediate, with the weight of evidence favouring the former.

9.1.4. *Aromatase Chemistry*

The P450 enzyme aromatase is remarkable for its catalysis of three successive steps at a single active site in the biotransformation of steroidal androgens through demethylation and concomitant aromatisation of the steroid A ring (*300–305*). For example, the metabolism of 4-androstene-3,17-dione into estrone commences with two hydroxylation reactions at the C-19 methyl group to generate an aldehyde, Fig. 18. The first monooxygenation in the aromatisation process is a hydroxylation for which the hydrogen abstraction-oxygen rebound mechanism discussed above should apply. The second monooxygenation can also be rationalised by a hydroxylation mechanism: hydrogen abstraction from the methylene carbon followed by oxygen rebound and dehydration of

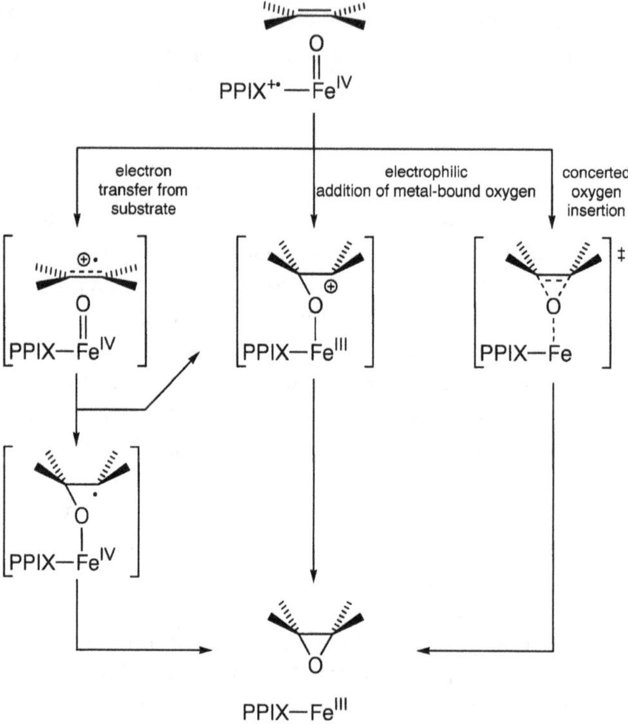

Fig. 17. Possible mechanisms of epoxidation reactions catalysed by cytochromes P450

the resulting aldehyde hydrate, although alternative pathways are feasible for this reaction (*300*). The third monooxygenation, an oxidative deformylation step which produces estrone and formate, is unusual and nucleophilic attack of a ferric heme peroxide species (PPIX-Fe(III)-OOH) on the aldehyde has been proposed to account for it. Again, alternative pathways are possible for this monooxygenation reaction, but the mechanism shown most directly accords with the results of isotopic labelling studies and has been favoured (*300*). Chemistry related to this latter reaction has been invoked in the NOS-mediated conversion of NHA into Cit (Section 9.3).

9.2. Mechanism of NOS Monooxygenation I

The mechanistic detail of the first NOS monooxygenation cycle represented in Fig. 19 is thought to feature reductive activation of O_2 at

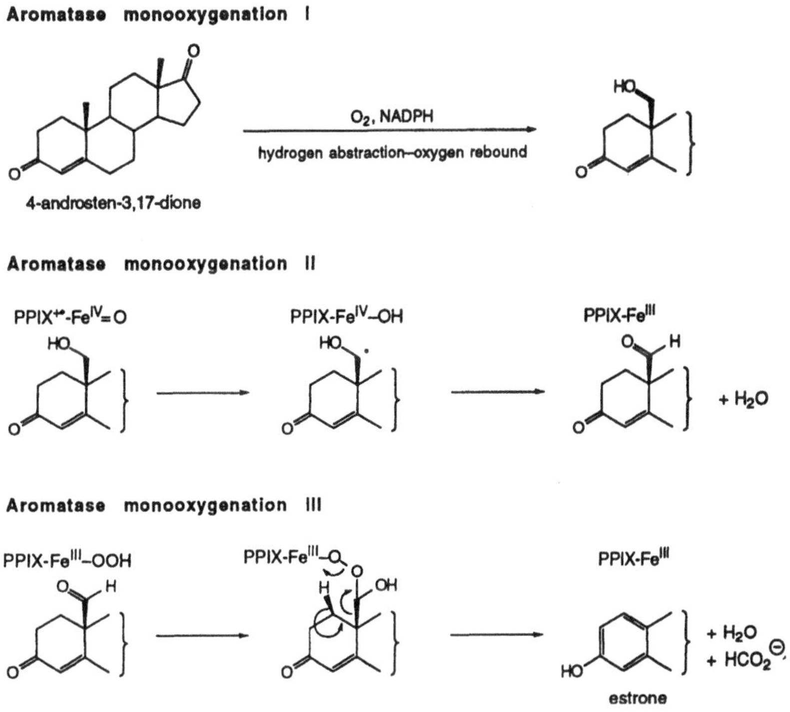

Aromatase monooxygenation I

Aromatase monooxygenation II

Aromatase monooxygenation III

Fig. 18. Aromatase-mediated biotransformation of androstenedione into estrone

the heme centre in the manner which is well precedented in cytochrome P450 chemistry. By analogy to cytochromes P450, then, the catalytic cycle might be initiated by substrate binding to the active site, possibly causing displacement of an axial ligand such as water from the iron (*Step A*). Thus, it is known for other cytochromes P450 that in the resting ferric state the enzyme exists in equilibrium between two forms: a penta-coordinate high-spin (S=2) iron complex and a hexacoordinate low-spin (S=0) complex (*129*). In the pentacoordinate state, the protein thiolate serves as the only axial ligand for the iron, the atomic diameter of which is too great for accommodation in the porphyrin central cavity, which has a typical radius of 2.01 Å (*306*). Consequently, in the pentacoordinate complex, the iron lies substantially out of the porphyrin plane towards the proximal thiolate ligand. The hexacoordinate low-spin iron, on the other hand, has a smaller atomic diameter and lies in the porphyrin plane; the sixth ligand is provided either by a water molecule or an alcohol residue from the protein. Substrate binding to the protein, close to the heme, is thought to shift the equilibrium in favour of the pentacoordinate high-spin state, which frequently exhibits a higher redox potential and is more readily reduced by NADPH. Reduction of the heme iron, with electron transfer from NADPH proceeding *via* the flavins in the NOS reductase domain, could produce a high-spin pentacoordinate ferrous complex (*Step B*). Dioxygen binding, *Step C*, would then lead to formation of a low-spin hexacoordinate oxyferrous complex.* A further one-electron reduction to afford a ferric peroxide adduct and protonation may ensue to give a hydroperoxide complex (*Steps D and E*); *Step D* is the rate-limiting step in the overall catalytic cycle of $P450_{cam}$. Protonation of the hydroperoxide might trigger heterolysis of the O–O bond, (*Step F*), and generate the transient high-valent oxoiron species for transfer of oxygen to the substrate. As discussed earlier, the precise electronic structure of this extremely short-lived intermediate has been the subject of much debate, and may be better formulated as an oxoiron(IV)protoporphyrin IX radical cation rather than an oxoiron(V)-porphyrin complex; it may also be represented by several other resonance canonical forms (*176*).

The formal protonation state of the enzyme-bound Arg guanidino group upon which the monooxygenation is effected has not been defined. It is represented here in the cationic state because recent work (*298, 299*) has demonstrated that the substrate is bound by a salt bridge between its guanidinium functionality and a glutamate residue in the catalytic centre. Mechanistic detail for the final oxygen transfer step, *Step G*, is not explicitly shown in Fig. 19; hydrogen abstraction from one of the two equivalent terminal guanidinium nitrogen atoms of the substrate to generate a nitrogen radical followed by oxygen rebound could, in principle, afford the *N*-hydroxylated intermediate NHA by a route consistent with conventional alkane hydroxylation (Fig. 20), and a number of authors have assumed such a mechanism to be operative (*308, 309*).

Marletta, however, has formulated the *N*-hydroxylation step of the NOS-catalysed reaction in terms of a mechanism involving an initial single electron oxidation of the substrate followed by oxygen rebound rather than a direct hydrogen abstraction from the guanidinium nitrogen, (Fig. 21) (*167*). This process is analogous to the single electron transfer from tertiary amines *en route* to *N*-oxides. Deprotonation of the substrate might in principle precede the electron transfer step and Marletta noted

---

* Binding of O_2 to PPIX-Fe^{II} is frequently represented as PPIX-Fe^{II} O_2, a complex in which no formal oxidation of the ferrous iron has occurred. It has been suggested, however, that this "end-on" complex of O_2 is better formulated as one between ferric iron and superoxide (*ie* PPIX-Fe^{III} O-O·) where electron transfer from iron to O_2 has formally occurred (*307*). The structure of this complex has been discussed in a recent review (*176*).

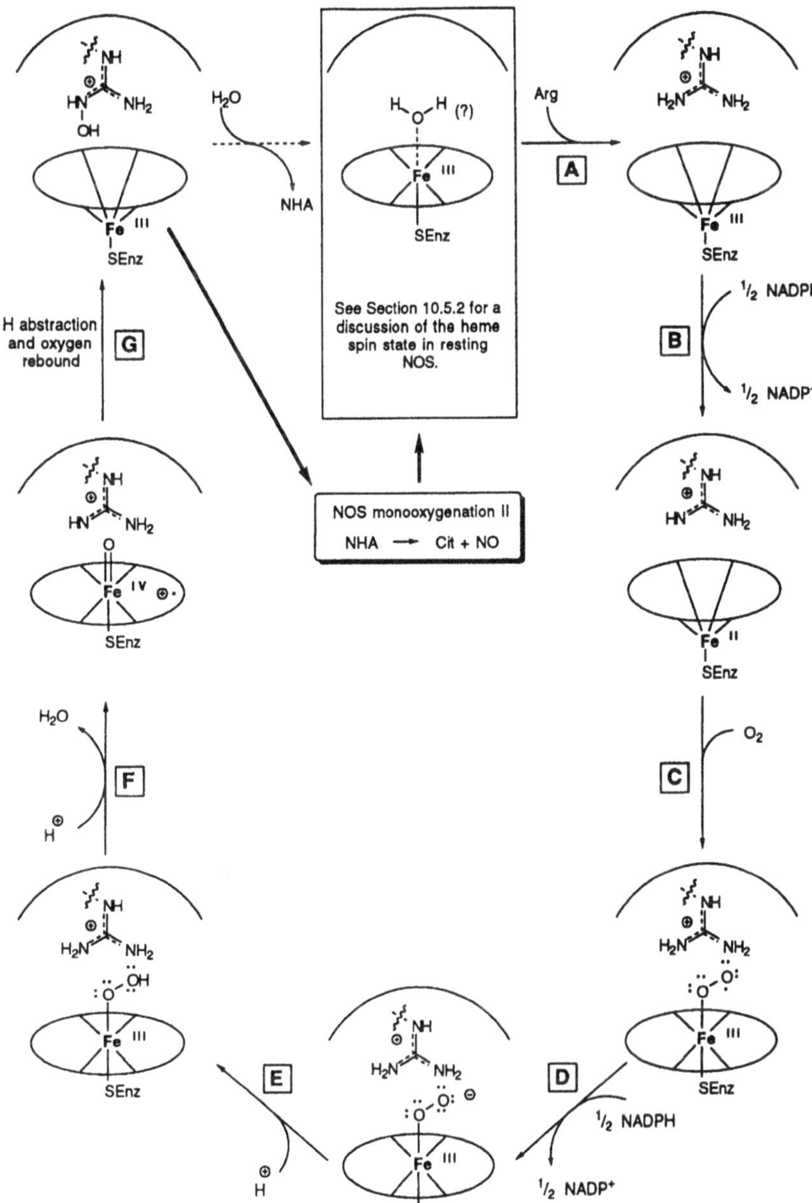

Fig. 19. Proposed catalytic cycle for the first monooxygenation reaction in the NOS-catalysed reaction

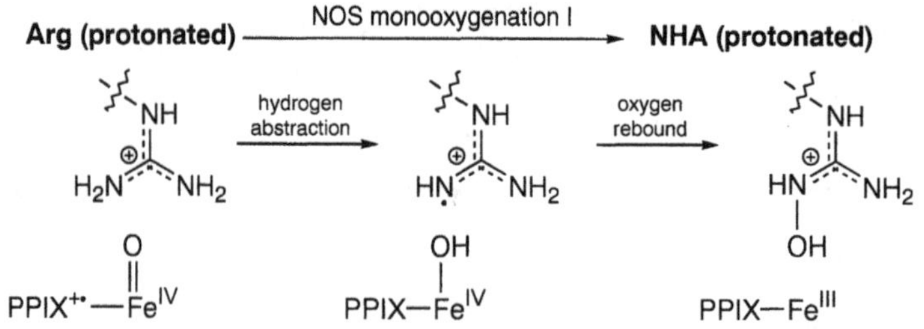

Fig. 20. Mechanistic detail for the first NOS-catalysed monooxygenation reaction proceeding by a hydrogen atom abstraction pathway

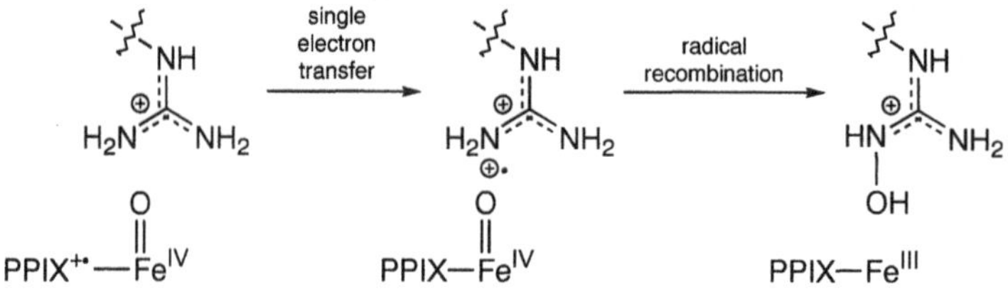

Fig. 21. Mechanistic detail for the first NOS-catalysed monooxygenation reaction proceeding by an initial single electron oxidation of the substrate as proposed by MARLETTA. Arg and NHA are shown in the protonated cationic state here, but deprotonation may possibly occur during or after NOS monooxygenation I

that the N-hydroxyguanidine product of this monooxygenation reaction is substantially less basic than the guanidine-containing substrate.

In contrast to the Marletta mechanism a recent quantum chemical modelling study of NOS monooxygenation I has prompted speculation that the reaction proceeds by a concerted oxygen insertion into the NH bond (*cf.* Fig. 20) mediated by the oxoiron intermediate (*310*). This mechanism is akin to the hydroxylation of alkanes by cytochromes P450. However, an initial two electron reduction of the Arg guanidinium moiety was postulated prior to the oxygen insertion step.

The epoxidation of alkenes and arenes is another well documented monooxygenation process mediated by the cytochromes P450. In the case of arene epoxidation, the reactive arene oxide intermediates usually rearrange to afford phenols. Although a P450-type N-oxidation mechanism is usually put forward to account for the first NOS monooxygenation, an "epoxidation" reaction might also be envisaged whereby oxygen insertion into the guanidine double bond would give rise to an oxaziridine moiety. Ring opening of this labile intermediate by the adjacent nitrogen atoms would also lead to formation of NHA, the intermediate in the NO biosynthetic pathway, Fig. 22. Indeed MARLETTA and co-workers have recently suggested such a pathway as a possible mechanistic

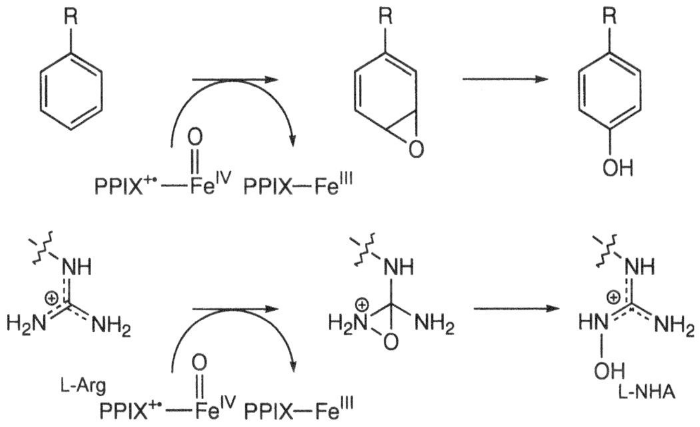

Fig. 22. Oxaziridine formation in a manner analogous to the epoxidation of alkenes and arenes could also account mechanistically for the NOS-catalysed oxidation of Arg to NHA

variation for the reaction (*311*). The primary feature of distinction between this mechanism and the direct *N*-oxidation mechanism discussed above is the C–O bond generated during the course of the *N*-hydroxylation pathway.

In summary, then, the first monooxygenation is thought to accord mechanistically with established cytochrome P450 *N*-oxidation chemistry in that it is probably, though not unequivocally, mediated by a transient, high-valent oxoiron porphyrin complex, and may commence with a single electron oxidation of the guanidinium (or guanidine) substrate followed by combination of the guanidinium radical with iron-bound oxygen. Hydrogen atom abstraction and oxygen rebound is an alternative mechanism to the electron transfer pathway which has also received consideration. Although the proposed chemistry of this monooxygenation is considered to be better defined than the second NOS monooxygenation of NHA to Cit, for which there appears to be no known precedent in biological chemistry, there remain unanswered questions relating to the fine mechanistic detail of the reaction. Substrate-binding may regulate the kinetics of electron entry facilitating reductive activation of dioxygen, and there is evidence to support this: in particular that substrate-binding does not trigger heme iron reduction, but does influence electron flux through the NOS redox centres (*312, 313*). The ease of electron transfer or hydrogen abstraction from the substrate will be a deciding issue in relation to the mechanistic detail of the reaction. In this regard, the positioning of substrate with respect to the perferryl oxygen in the catalytic site as well as the protonation state of the guanidine will have a major impact, but little is known about the structure of the catalytic site as yet.* An additional feature that requires clarification is the nature of the involvement in NOS catalysis of the biopterin, which is located close to the heme and substrate binding sites. Throughout the above discussion it has been assumed that the hydroxylation step is catalysed by the action of heme without the direct participation of H_4B as a redox active cofactor. This is generally accepted to be the case although it has not yet been unequivocally proven (*311*). The possible function of H_4B in NOS catalysis is reviewed in Section 10.6.

9.3. Mechanism of NOS Monooxygenation II

The mechanistic detail of the second NOS monooxygenation (NHA → Cit + NO) has been debated extensively in the literature. Again, mechanisms put forward for this transformation in recent years have featured heme-mediated monooxygenation rather than direct involvement of H_4B as a redox active cofactor. MARLETTA initially proposed a pathway in which NHA was oxidised to a *C*-nitroso derivative (Fig. 23) (*314, 315*). Hydrogen abstraction from this intermediate (or single electron oxidation and deprotonation) followed by homolytic

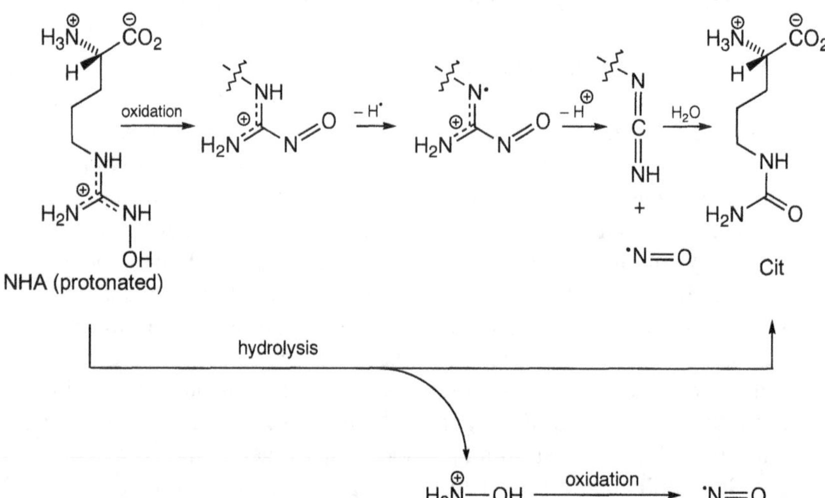

Fig. 23. Early mechanisms for the second NOS monooxygenation put forward by MARLETTA in 1988 (*314, 315*) and DEMASTER *et al.* in 1989 (*317*)

---

* See Addendum.

fragmentation would afford NO and a carbodiimide; Cit would issue from hydration of the latter. An alternative mechanism was also postulated at an early stage in which hydrolysis of NHA would afford Cit and hydroxylamine; enzyme-mediated oxidation of the latter species, a known process in some bacteria (*316*), would then generate NO (*317*).

With a view to shedding further light on this unusual reaction FUKUTO and co-workers examined the chemical oxidation of model guanidoximes such as *N*-(*N*-hydroxyamidino)piperidine (NHAP) (*120*). Several agents were found capable of oxidising the guanidoxime function (Fig. 24), some producing significant amounts of NO but others generating N_2O. Formation of the latter arises from initial oxidation to HNO, which is known to dimerise to hyponitrous acid and decompose to water and N_2O extremely rapidly in aqueous solution. In each case the final organic product formed, however, was not a urea but a cyanamide *via* a presumed nitroso intermediate analogous to that proposed by MARLETTA (Fig. 23) in the NOS-mediated transformation of NHA. Other nonenzymatic oxidative denitrification pathways have also been studied in attempts to model the formation of Cit and NO from NHA. SENNEQUIER and co-workers, for example, have found that superoxide efficiently cleaves guanidoximes with formation of nitrogen oxides (*318*). EVERETT *et al.* have also studied the oxidation of NHA both by superoxide and by azide radical (*319*). NO was formed in both cases, and the reaction with superoxide, in principle, might occur *in vivo* (*320, 321*), although NOS-dependence in the biological transformation of NHA into Cit is unequivocally established.

The mechanisms proposed in Fig. 23, in which the ureido oxygen atom of Cit originates from water, were excluded when it was shown by

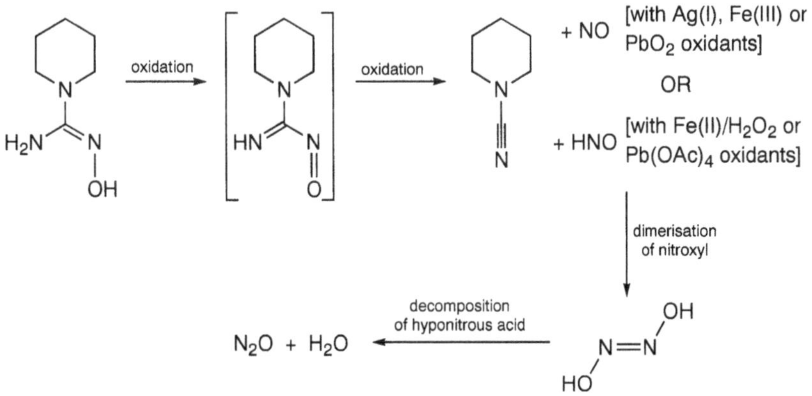

Fig. 24. Chemical oxidation of model *N*-hydroxyguanidines such as *N*-(*N*-hydroxyamidino)piperidine were studied by FUKUTO *et al.* (*120*) in the period up to mid-1991

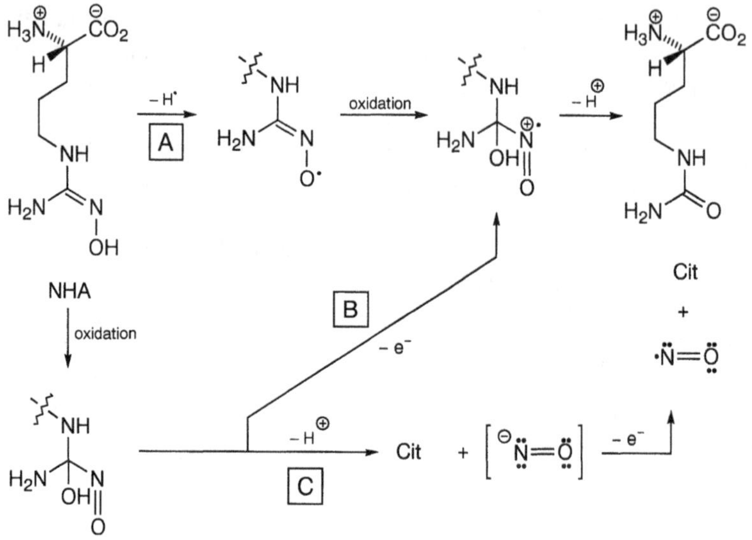

Fig. 25. Mechanisms postulated by GRIFFITH and STUEHR (*323*) in 1992 to account for the second NOS monooxygenation

isotopic labelling studies that the Cit oxygen is derived exclusively from O_2 (*322*). With this knowledge STUEHR and GRIFFITH proposed three possible pathways for the oxidation of NHA in 1992, Fig. 25 (*323*).

In pathway A, hydrogen abstraction followed by *C*-hydroxylation of the guanidine to afford an intermediate radical cation was postulated; deprotonation and fragmentation would complete the pathway to Cit and NO. In pathways B and C the hydrogen abstraction/*C*-hydroxylation steps are reversed. Thus *C*-hydroxylation of the guanidine would afford an intermediate nitroso compound which might (*via* pathway B) undergo single electron oxidation to afford the same radical cation intermediate proposed in pathway A or alternatively fragment directly to Cit and NO^-, the latter being oxidised subsequently to NO. Each of the three pathways involves a hydroxylation of the guanidino carbon atom (formal attack of HO^+) and at some stage, either before or after this step, an additional one-electron oxidation of the substrate. STUEHR and GRIFFITH further suggested that the *C*-oxygenation process might involve H_4B, Fig. 26, in view of the role of this cofactor in monooxygenations catalysed by H_4B-dependent aromatic amino acid hydroxylases (see Section 10.6). Although H_4B-mediated hydroxylation at the guanidino carbon atom is unprecedented, nucleophilic attack of a 4*a*-hydroperoxy-H_4B species on the electrophilic carbon atom of NHA (or its iminoxyl radical derivative) was supported by theoretical studies which indicated

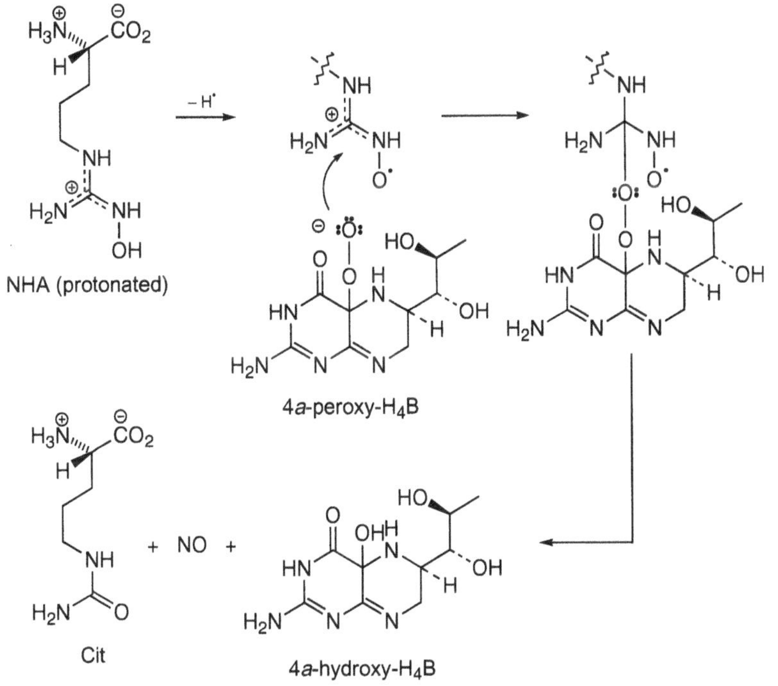

Fig. 26. Early mechanistic detail postulated by Griffith and Stuehr in 1992 to account for the second NOS monooxygenation: nucleophilic attack by the H_4B-derived hydroperoxy species generates a tetrahedral peroxy adduct that rearranges to Cit and NO

that the carbon atom of an N-hydroxyguanidinium species is electron deficient in character (*324*, *325*) and because NHA has been found susceptible to attack by nucleophiles such as hydroxide at this position (*326*).

A variation on the proposed NOS monooxygenation II mechanism was advanced later in 1992 featuring an initial single electron oxidation of the guanidoxime function rather than a hydrogen atom abstraction (Fig. 27) (*327*). With the discovery of NOS heme-dependence it was also suggested that the putative nucleophilic peroxy oxidant might be derived from heme as an alternative to H_4B.

Fukuto *et al.* provided further indirect evidence supporting the involvement of a peroxide oxidant in NOS monooxygenation II from examination of the reaction between peracids and guanidoxime models for NHA (*121*). The reaction of NHAP with *meta*-chloroperbenzoic acid (*m*CPBA), for example, was found to generate HNO and a urea product, Fig. 28. In this respect the reaction resembled the NOS-mediated

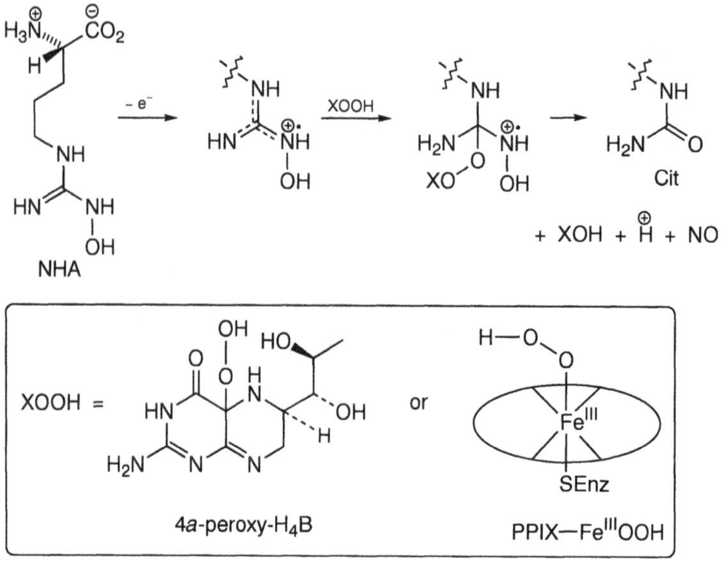

Fig. 27. By 1993 a mechanism involving nucleophilic attack of a NOS-bound cofactor peroxide on the guanidoxime C atom of NHA had been proposed for the NOS monooxygenation II pathway

transformation of Arg into the urea product Cit, with the ureido oxygen atom being derived from the oxidant rather than water. A mechanistic scheme was advanced for the model reaction which involved stepwise formation of an oxaziridine intermediate following nucleophilic attack by the peracid on the guanidoxime carbon atom (Fig. 28).

The metastable HNO formed in the reaction of NHAP with mCPBA was detected indirectly through formation of its stable decomposition product N_2O. Formation of NO itself was *not* observed and would require an additional single electron oxidation at some point in the reaction mechanism. This additional oxidation might occur either before or after the postulated nucleophilic attack of the enzyme-derived peroxide in the NOS monooxygenation. FUKUTO *et al.* initially represented the extra oxidation, in preceding the nucleophilic attack, as a single electron transfer from the substrate followed by an eventual deprotonation (pathway A Fig. 29). In succeeding the nucleophilic attack, the additional single electron oxidation might occur at the stage of one of the metabolic intermediates, such as an oxaziridine (pathway B), or at the end of the mechanism by oxidation of HNO (pathway C).

The nature of the postulated enzyme-derived peroxy nucleophile remained an important issue to address. Although a function for $4a$-hydroperoxy-H_4B in this role – as originally suggested – could not be

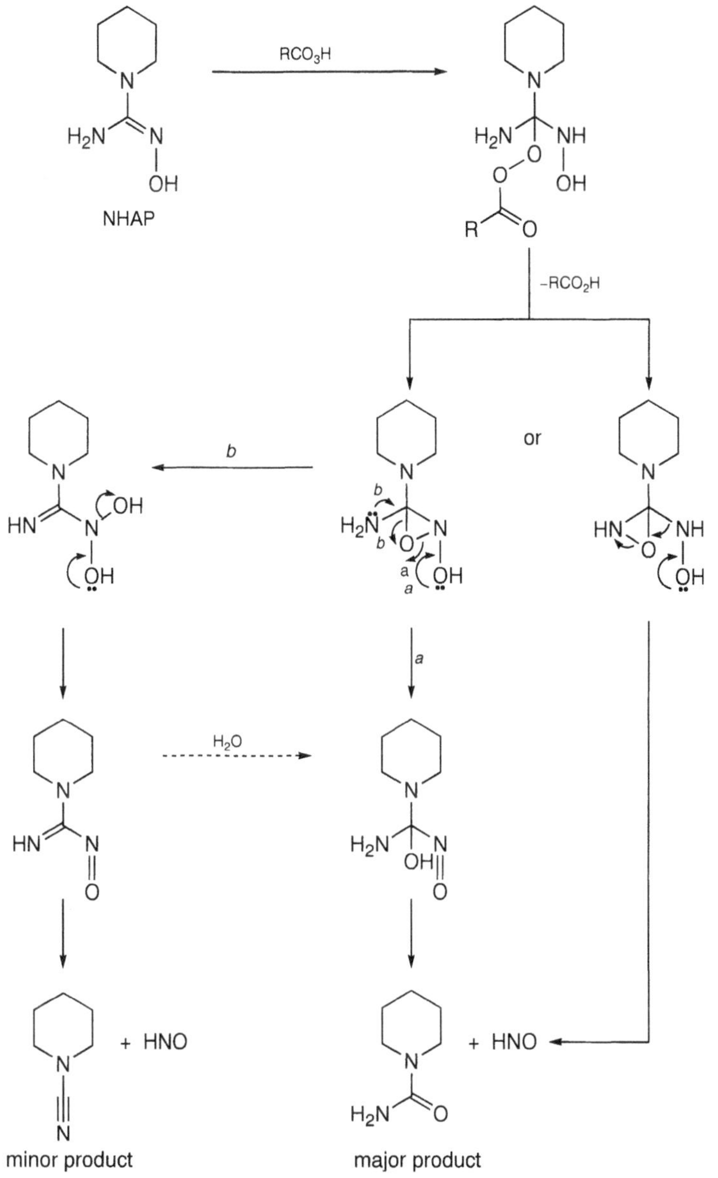

Fig. 28. Possible reaction pathways for the peracid-mediated oxidation
of *N*-(*N*-hydroxyamidino)piperidine

excluded, the relation of NOS to the cytochrome P450 family and
emerging studies showing that P450s can also oxidise amidoximes (*328*)
and NHA (*329, 330*) to NO strongly implicated a heme-derived peroxy

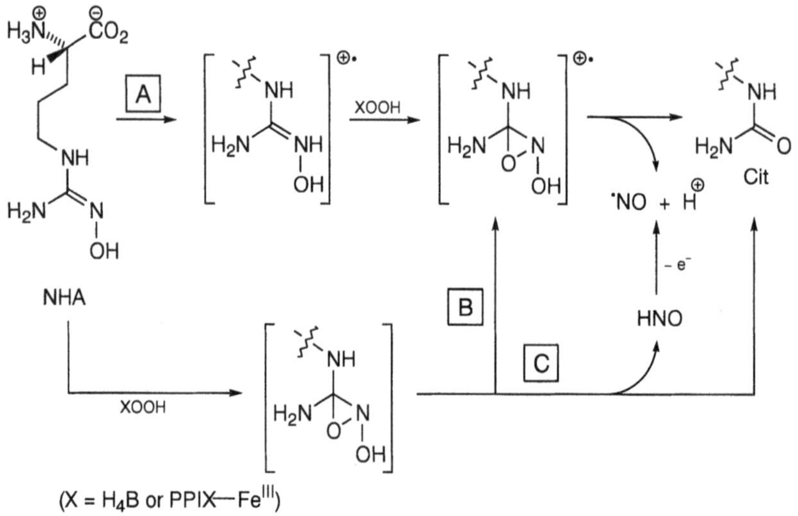

Fig. 29. Possible mechanistic pathways suggested for the second NOS monooxygenation in the light of the model oxidation reactions with *N*-hydroxyguanidines

species in this role. The known reactions of other iron-peroxo species at electrophilic centres was also cited in support of a role for the heme peroxide complex as the attacking nucleophile in NOS monooxygenation II (*331*). Thus, it was speculated that enzyme-bound heme mediates *both* NOS monooxygenations I and II. Interestingly, Clement and co-workers have shown subsequently that cytochrome P450 systems other than NOS can also facilitate *N*-hydroxylation of guanidines and *N*-aminoguanidines in reactions that parallel the first NOS monooxygenation (*332, 333*). Studies by a number of groups have shown that CO inhibits both NOS monooxygenations and strengthen the case for heme involvement in these reaction pathways (*136, 139, 334, 336*).

A key review by Marletta in 1993 further advanced ideas concerning the mechanism of the second NOS monooxygenation (*167*). The proposed pathway featured nucleophilic attack on the guanidine carbon by a ferric heme peroxide complex (PPIX–Fe(III)–OO⁻), with the latter being formed from the ferric heme superoxide complex (PPIX–Fe(III)–OO·) by hydrogen abstraction or single electron oxidation and deprotonation of the substrate hydroxy group. Marletta proposed a direct fragmentation of the peroxide-substrate adduct to Cit and NO rather than invoking formation of oxaziridine intermediates (see Fig. 29) though mechanistic variants involving the intermediacy of such species were also envisaged. Feldman, Griffith and Stuehr considered a mechanistic variation of this scheme in which the initiating step was explicitly formulated as a single electron oxidation rather than a hydrogen abstraction (*336*). It was subsequently argued by KORTH and co-workers (*307*), however, that the redox potential of the ferric heme superoxide complex was unlikely to be sufficiently high as to favour a single electron transfer from the substrate to generate the radical cation of NHA. A mechanistic scheme was presented (Fig. 30), in which the substrate suffers the hydrogen abstraction and nucleophilic attack proposed by MARLETTA.* ISHIKAWA *et al.* have recently proposed a chemical model for aspects of this biological transformation based on a photo-sensitised oxygenation of a guanidoxime model (*338*).

The involvement of PPIX–Fe(III)–OOH proposed in the second NOS monooxygenation has parallels with the known chemistry of aromatase and related P450 enzymes (*300, 301*). Aromatase effects three successive steps leading to oxidative demethylation of steroidal androgens (Section 9.1.4 and Fig. 18). Nucleophilic attack of ferric heme peroxide on a steroidal aldehyde intermediate has been invoked as a key step in this process and provides some precedent for the chemistry proposed in the NOS-mediated conversion of NHA into Cit. It is uncertain why attack of PPIX–Fe(III)–OOH on the aldehydic substrate should precede scission of the peroxide bond that would generate the high-valent oxoiron species which features so prominently in P450 chemistry.[†] MARLETTA has suggested that the electrophilicity of the aldehyde might dictate the reaction course and facilitate nucleophilic attack before O-O bond scission to $PPIX^{+\cdot}$–Fe(IV)=O occurs (*167*). Similarly, FELDMAN *et al.* have argued that the known susceptibility of NHA to nucleophilic attack, for example by hydroxide or ammonia to produce Cit and Arg respectively, indirectly supports the role proposed for the nucleophilic oxidant in the NOS mechanism (*336*).

---

* At the point of oxidation the precise protonation state of NHA, which is more stable in its hydroxyimino tautomeric form, is uncertain. The imino nitrogen is expected to constitute the normal site of protonation in guanidine structures due to resonance stabilisation of the positive charge in the conjugate acid. The influence of the electron withdrawing hydroxyl group, however, renders *N*-hydroxyguanidines substantially less basic than the corresponding parent guanidine structures. The pK_a value of *N*-hydroxy-guanidine is close to 8.0, for example, which compares with a value of 13.6 for guanidine; other substituted guanidoximes have pK_a values in the range 7.7–8.4 (*721, 337*). Consequently, the precise protonation state of the substrate in the second NOS monooxygena-tion will require clarification. For simplicity of representation we have shown the substrate in an unprotonated form.

[†] A wider role for peroxo-iron as an alternative oxygenating species to oxenoid-iron has recently been proposed in P450-catalysed biological chemistry (*339*).

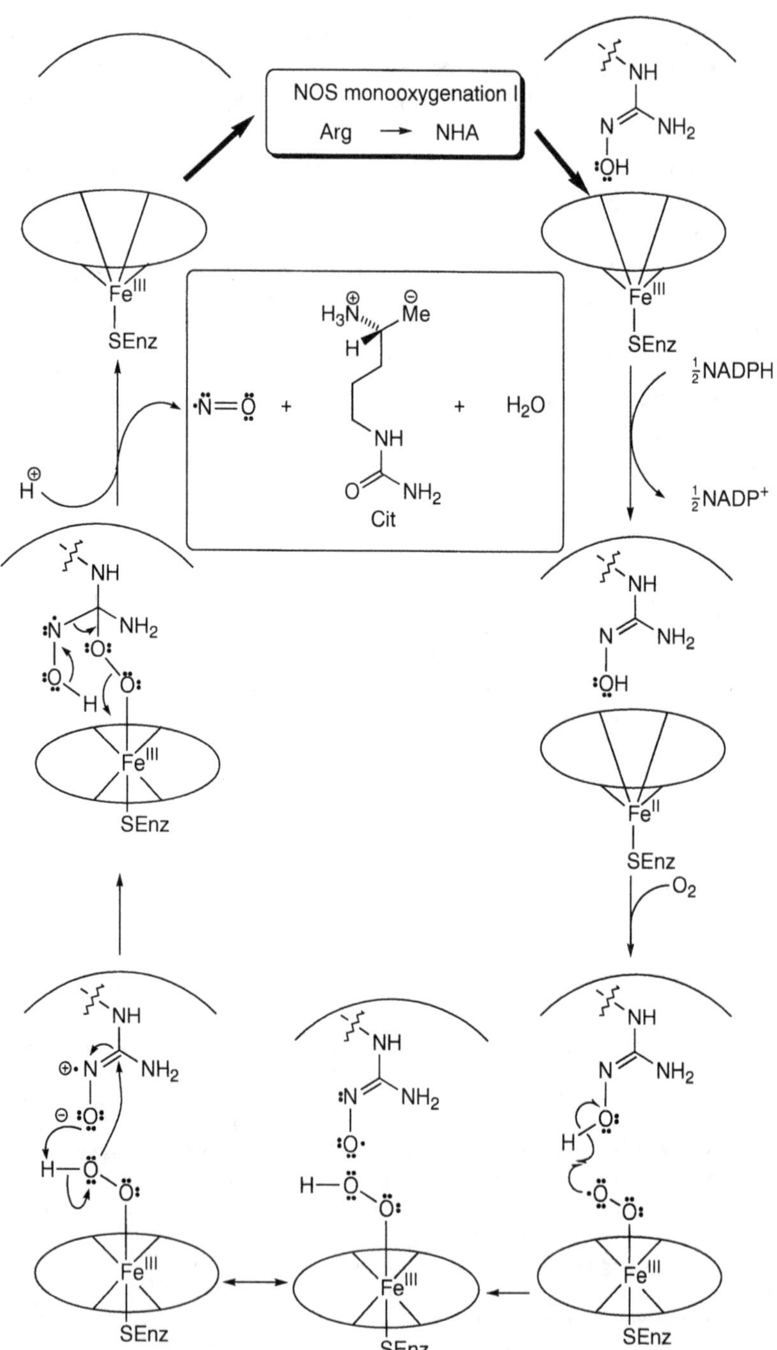

Fig. 30. Mechanistic detail for the second NOS monooxygenation

MARLETTA and co-workers have recently attempted to probe the NOS mechanism in oxygen donor-supported incubations of the murine macrophage isoform with Arg and NHA (*311*). Oxygen donor surrogates such as organic hydroperoxides, H_2O_2 and PhIO are known to substitute for NADPH and O_2 in reactions catalysed by cytochromes P450 (*340*). These compounds, by means of the peroxide shunt pathway (Fig. 31), give the same oxoiron intermediate that is formed by reductive activation of O_2. In the case of the single oxygen atom donor PhIO the intermediate is formed directly whereas from the peroxides it is presumably formed stepwise *via* a ferric heme peroxide species. As the oxoiron intermediate is thought to mediate NOS monooxygenation I but the iron peroxide to effect NOS monooxygenation II, the reaction with PhIO was expected to facilitate only the formation of NHA from Arg while H_2O_2 might, in principle, support both reactions.

In practice, the first monooxygenation was found to take place neither with PhIO, H_2O_2 nor with alkylhydroperoxides upon incubation of Arg with NOS. In contrast conversion of NHA into Cit was observed in H_2O_2-supported NOS incubations, though not those supported by alkyl hydroperoxides. Use of PhIO was complicated by the fact that

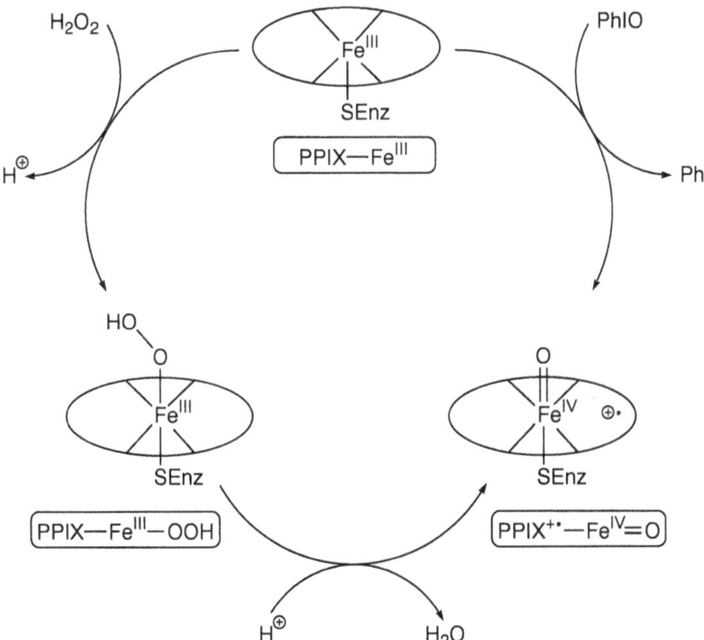

Fig. 31. The peroxide shunt pathway

NHA reacts non-enzymatically with this oxidant to produce NO_2^-/NO_3^- but enzymatic assays were not found to produce NO_x^- above these non-enzymatic levels. The failure of all the oxygen donor surrogates to support the first monooxygenation may be due to: (i) the difficulty of lipophilic oxygen donors (PhIO, ROOH) to access a potentially hydrophilic active site, and (ii) the formation of a ferric peroxide complex from H_2O_2 but failure of O–O bond scission leading to the oxoiron intermediate. Alternatively, if the latter intermediate is actually formed then the result would suggest that the first monooxygenation is not effected by heme but might possibly involve H_4B, an option that cannot be excluded at present.

Results of the NHA incubations were consistent with the second NOS monooxygenation as currently proposed. It is predicted in this case that the co-product of Cit in this reaction should be NO^- rather than NO because of the absence of the ferric superoxide complex which is supposed to mediate the one-electron oxidation of the substrate in the $NADPH/O_2$-supported pathway (Fig. 30). Indeed formation NO^- was inferred in these studies through the detection of its more stable decomposition product N_2O. However, the small quantity of N_2O formed, based on that expected for stoichiometric conversion of NHA into Cit and NO^-, suggested that the majority of nitroxyl decomposes aerobically and in the presence of H_2O_2 to form NO_x^-.

9.4. Mechanism-Based Inhibitors of NOS

The NO synthases, though possessing very high substrate specificity, also mediate the transformation of certain Arg derivatives such as N^G-monomethyl-L-arginine (NMA), a mechanism-based inhibitor which is slowly processed by the enzymes (Fig. 32). MARLETTA and OLKEN suggested that NOS processing of NMA begins with single electron oxidation of the substrate to generate a guanidinium radical (*341–344*). Oxygen rebound and deprotonation of the latter would afford the *N*-hydroxylation product, N^G-hydroxy-N^G-methyl-L-arginine (NHMA), by the same NOS monooxygenation I mechanism proposed for conversion of Arg into NHA. On the other hand a competing *C*-deprotonation of the guanidinium radical followed by oxygen rebound would lead, *via* α-amino radical and carbinolamine intermediates, to Arg and formaldehyde according to the type B pathway for P450-mediated *N*-dealkylations discussed in Section 9.1.2. Further oxidative processing of NHMA might also generate formaldehyde *via* the hydrolysis of an intermediate nitrone, forming NHA in so doing. Arg and NHA once formed would of

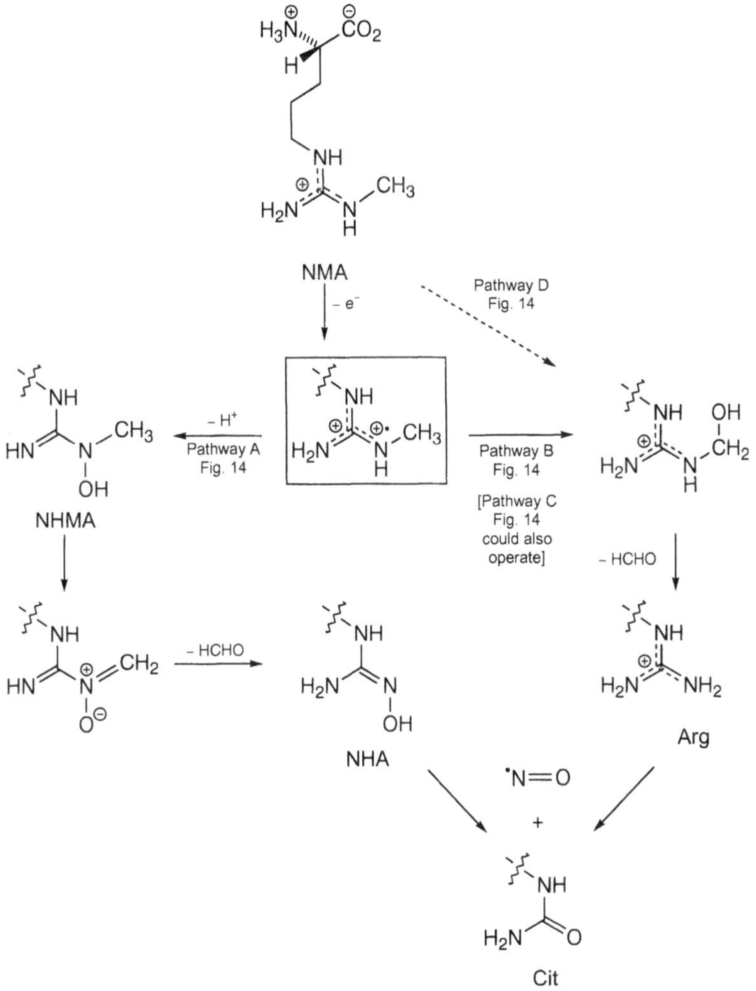

Fig. 32. Mechanism proposed by Marletta and Olken (*341–344*) to account for NOS-catalysed processing of N^G-monomethyl-L-arginine (NMA)

course be processed to NO and Cit by the normal NOS-catalysed pathway. Significantly, NHMA, Arg and formaldehyde were all detected during NOS processing of NMA; chemically synthesised NHMA was also shown to form Cit upon incubation with NOS (*343*).

Interestingly, Olken and Marletta's results showed that NMA was a *partially uncoupled*, slow alternative substrate for NOS; that is, a significant amount of H_2O_2 is generated upon NOS incubation with

NMA due to uncoupling of the O_2 activation and substrate oxidation processes (*343*). It was estimated that approximately 70% of the NADPH consumed under these assay conditions resulted in H_2O_2 formation. This contrasts markedly with NOS turnover of Arg where substrate oxidation is relatively tightly coupled to the process of O_2 activation except for when the medium is depleted (<10 µM) in Arg (*345*). Since the formation of H_2O_2 is known (*346, 347*) to inactivate other P450 enzymes through oxidative heme destruction the question arose as to whether H_2O_2 formation during NOS turnover of NMA might bring about irreversible enzyme inhibition. However, neither H_2O_2, HCHO nor NO formation from NMA or NHMA was found to contribute significantly to NOS inactivation at the concentrations generated in these investigations. Instead it was postulated that irreversible NOS inhibition by NMA might be caused by the reactivity of the α-amino radical invoked in *N*-dealkylation of NMA through covalent modification of the heme or apoprotein in a proportion of substrate turnovers. Alternatively the electrophilic nature of a nitrone intermediate or the iminium ion arising from collapse of the putative carbinolamine might also lead to enzyme inactivation by reaction with nucleophilic centres on the protein.

In related work involving other NOS inhibitors, N^G-allyl-L-arginine (NALA) and N^G-cyclopropyl-L-arginine (NCPA), attempts were made by Marletta and Olken to probe further the possible *N*-dealkylation pathways and mechanism of NOS inhibition by N^ω-substituted L-arginines (*342*). An initial hydrogen abstraction at the α-carbon ought to be promoted in the case of NALA due to the ease of oxidation at an allylic methylene, although single electron oxidation followed by *C*-deprotonation would give rise to the same allylic radical (Fig. 33). Single electron transfer from the guanidinium nitrogen atom of NCPA, on the other hand, was expected to lead to rapid rearrangement of the intermediate nitrogen radical cation to generate a carbon-centred radical, causing irreversible inhibition of the enzyme by its reaction with protein residues in the catalytic site or with the heme prosthetic group. In the event NALA showed irreversible inhibition of NOS whereas NCPA inhibited the enzyme only reversibly (*i.e.* by competition for the Arg binding site). However, these observations do not conclusively rule out single electron oxidation as the initial step in the process since NCPA, which is sterically more demanding and less flexible than NALA, may undergo an alternative form of processing by NOS or simply fail to serve as a substrate. As discussed earlier evidence for a single electron oxidation as the first step in *N*-oxygenation reactions by cytochromes P450 is convincing (*264*) and also thought likely to occur in the formation of NHA from Arg catalysed by NOS.

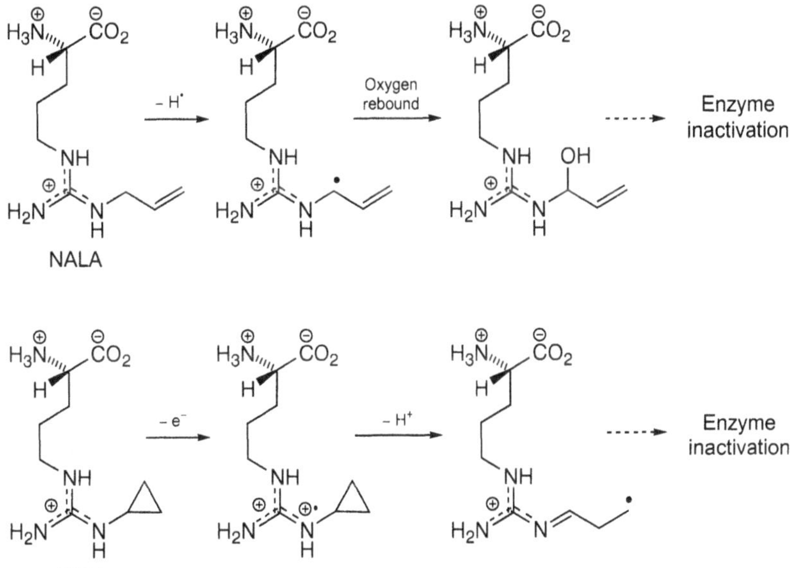

Fig. 33. Possible NOS-catalysed processing of N^G-allyl-L-arginine (NALA) and N^G-cyclopropyl-L-arginine (NCPA)

An alternative pathway for the processing of NMA was suggested by FELDMAN *et al.* in 1993 (*327*). Like the Marletta and Olken scheme it was proposed that transformation of the substrate commences in a manner analogous to NOS monooxygenation I, that is, with *N*-hydroxylation mediated by NOS heme occurring specifically at the methyl-bearing nitrogen to afford NHMA. However, further processing of NHMA was postulated by a mechanism common to that involved in the oxidation of NHA to Cit, (See Fig. 8) rather than by *N*-demethylation reactions. Such a pathway might be expected to generate Cit and a highly reactive nitrosomethane radical cation ($O{=}N^{+\cdot}{-}CH_3$) as the proximal products. This mechanism was formulated by an initial single electron oxidation of the guanidoxime function followed by nucleophilic attack of a cofactor-derived peroxide moiety on the guanidine carbon atom. At the time it was proposed that either 4*a*-hydroperoxy-H_4B or the ferric heme peroxide complex might constitute the attacking peroxide in this mechanism. FELDMAN *et al.* were unable to detect the formation of $O{=}N^{+\cdot}{-}CH_3$ through observation, for example, of a nitrosomethane-heme complex which might have accounted for the irreversible inactivation of the enzyme brought about by processing of NMA. Moreover, the mechanism does not account for MARLETTA'S observation of formaldehyde production upon NOS processing of NMA.

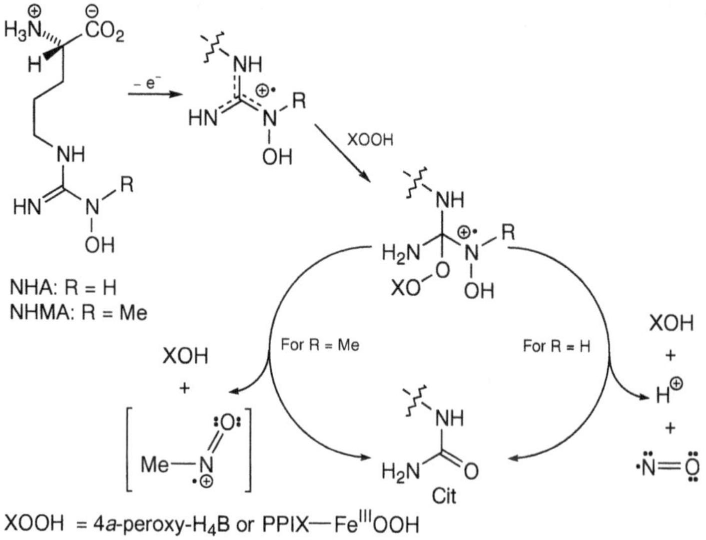

NHA: R = H
NHMA: R = Me

For R = Me

For R = H

XOOH = 4a-peroxy-H_4B or PPIX—$Fe^{III}OOH$

Fig. 34. Reaction pathway proposed by FELDMAN and associates (*327*) for NOS processing of NMA *via* NHMA

In further studies with NMA MARLETTA and co-workers have conclusively demonstrated that substantial loss of the heme chromophore accompanies enzyme inactivation (*348*). No single major heme adduct could be detected in these studies, however, and the authors proposed that multiple modes of inactivation are probable. Radiolabelling experiments suggested that a significant amount of NOS modification occurs by incorporation of a species derived from the *N*-methylguanidine moiety but lacking the amino acid backbone. This observation is consistent with the formation of the nitrosomethane radical cation.

9.5. Summary

Two quite different mechanisms, then, have been proposed for NOS monooxygenations I and II which transform Arg into NHA and then NHA into Cit and NO. It is generally accepted that both monooxygenations are mediated by activation of O_2 at the heme centre but several issues still require clarification; these include the precise function of H_4B in the mechanism, the protonation state of Arg and NHA during the reaction course, the timing of possible proton transfers and a satisfactory explanation for the participation of oxenoid-iron chemistry in monooxygenation I but peroxo-iron chemistry in monooxygenation II.

The state of substrate protonation requires clarification because of its likely impact on the single electron oxidation potential of the substrate in NOS monooxygenation I and because of its influence on the susceptibility of the guanidoxime function towards nucleophilic attack by PPIX–Fe(III)–OOH in NOS monooxygenation II. Guanidoximes are significantly less basic than their guanidine counterparts. Moreover, a self-consistent field molecular orbital study of *N*-hydroxyguanidine suggests that dissociation of the *N*-hydroxyguanidinium ion in neutral or alkaline media may occur by loss of the hydroxyl proton as well as by *N*-deprotonation (Fig. 35) (*324*).

The oxoiron and ferric peroxide oxidants that are supposed to mediate the NOS monooxygenations according to the prevailing mechanisms are fundamentally distinct from one another; the former is electrophilic and the latter nucleophilic in character. Operation of oxenoid-iron chemistry versus peroxo-iron chemistry therefore requires explanation. For example, nucleophilic attack by PPIX–Fe(III)–OOH on Arg, in a manner paralleling that proposed for its reaction with NHA, could be invoked in NOS monooxygenation I by interposition of an oxaziridine in the reaction pathway (Fig. 36). A number of factors may

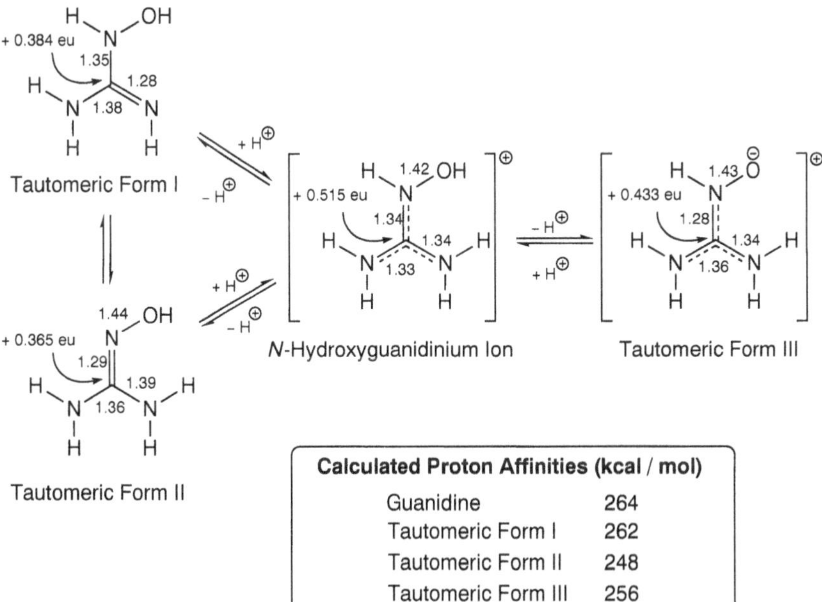

Calculated Proton Affinities (kcal / mol)	
Guanidine	264
Tautomeric Form I	262
Tautomeric Form II	248
Tautomeric Form III	256

Fig. 35. Results of molecular orbital calculations for *N*-hydroxyguanidine species (*324*). Calculated bond lengths (Å) are marked on the structures; net atomic charges (electron units) are indicated for the central guanidine carbon atom

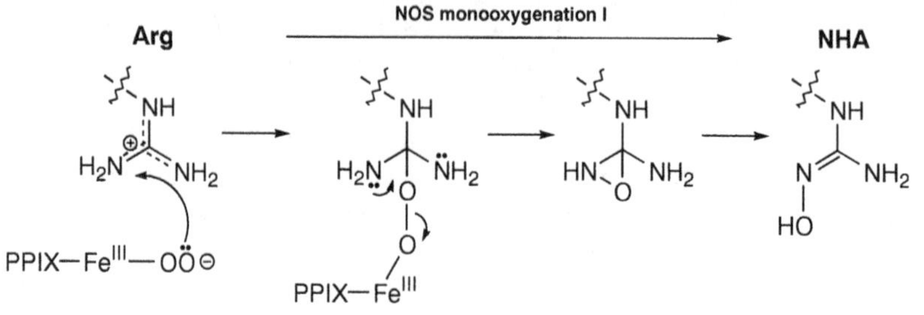

NOS monooxygenation I

Arg   NHA

$PPIX—Fe^{III}—O\ddot{O}^{\ominus}$

$PPIX—Fe^{III}$

Fig. 36. Nucleophilic attack of enzyme-bound ferric heme peroxide could account mechanistically for NOS monooxygenation I

potentially contribute to the operation of divergent reaction mechanisms in the NOS monooxygenations, if indeed current mechanistic formulations are correct. The involvement of a peroxo-iron oxidant in monooxygenation II but not in the oxidation of Arg may simply reflect a greater susceptibility of the guanidoxime function towards nucleophilic attack. However, the delivery of protons to the catalytic centre may also conceivably influence the reaction course in that peroxide bond cleavage ($PPIX–Fe(III)–OOH \rightarrow PPIX^{+\cdot}–Fe(IV)=O$) proceeds with consumption of a proton equivalent to generate the oxoiron species. Recent evidence by Vaz *et al.* certainly supports the idea that peroxo-iron chemistry is enhanced at the expense of oxenoid-iron chemistry if the internal proton transport chain within a P450 protein is disrupted by mutation of key hydrophilic residues (*339*). Thus conformational differences between NOS–Arg and NOS–NHA complexes that reduce the availability of protons to the ferric peroxide intermediate could dictate the reaction mechanism. Moreover, the substrate guanidinium ion itself may potentially supply a proton for scission of the peroxide bond in the first monooxygenation which would lead to formation of the less basic guanidoxime function of NHA in an unprotonated form.

Another important issue relating to the divergent mechanisms proposed for NOS monooxygenations I and II concerns the reduction of the ferric superoxide complex during O_2 activation. In the first monooxygenation reduction of $PPIX–Fe(III)–OO^{\cdot}$ is thought to proceed by electron transfer from NADPH whereas reduction by hydrogen abstraction from the substrate has been proposed for the second monooxygenation pathway. If the latter reaction were also to involve NADPH–sponsored reduction of $PPIX–Fe(III)–OO^{\cdot}$ to $PPIX–Fe(III)–OO^{-}$ then the immediate oxidation product ought to be nitroxyl rather than NO. In the case of $P450_{cam}$ the rate-limiting step for the overall catalytic cycle is the introduction of the second electron (*176, 287, 349*). Thus rate-limiting electron transfer from NADPH *via* the flavins might account for the preferred hydrogen abstraction pathway postulated in NOS monooxygenation II, and FUKUTO has commented in a recent review (*350*) on the propensity of NHA to act as a hydrogen donor agent. It is further conceivable that the nature of the enzyme-bound substrate, Arg (probably protonated) versus NHA (possibly unprotonated), sufficiently perturbs the heme environment so as to influence the redox potential of the ferric superoxide complex, thereby exerting control over electron flux from NADPH and thus over the source of reduction for the heme complex.

As earlier discussed reductive activation of O_2 in NOS processing of the inhibitor NMA is partially uncoupled from substrate oxidation and leads to production of a significant number of H_2O_2 equivalents per substrate turnover. Oxidation of Arg, however, is more efficiently coupled to reductive activation of O_2. An explanation for this difference in behaviour may be forthcoming as a clearer picture of the active site topology emerges. It is also interesting that the inhibitor, although initially processed to NHMA in a manner consistent with the first NOS monooxygenation mechanism, could be further transformed by pathways involving *N*-demethylation rather than by the hydroxy hydrogen abstraction and ferric heme peroxide attack proposed for NOS monooxygenation II. This is important because the *N*-demethylation mechanisms should be mediated by oxenoid-rather than peroxo-iron chemistry and, therefore, the presence of a guanidoxime substrate does not appear to preclude oxenoid-iron chemistry at the NOS heme centre. Orientational differences of the substrate that favour hydrogen abstraction from the hydroxy group and nucleophilic attack by PPIX–Fe(III)–OO^- might explain why processing of NHA does not involve the oxoiron intermediate, if indeed the prevailing mechanism for NOS monooxygenation II is actually operative.

The spatial and orientational arrangement of the substrate with respect to the heme catalytic centre is clearly a feature that requires investigation, and particularly the degree of reorientation that is required to accommodate two such different reaction mechanisms as are currently proposed for the NOS monooxygenations. The active site of aromatase, apparently, is able to catalyse different reaction types, a conventional hydroxylation mediated by oxenoid-iron and an oxidative deformylation thought to be effected by peroxo-iron. However, it is possible to unify the mechanisms of NOS monooxygenations I and II by invoking the intermediacy of oxaziridines as summarised in Fig. 37. Thus attack of the oxoiron complex on both Arg and NHA can be envisaged to generate labile oxaziridines by analogy to P450-mediated epoxidations. In the absence of coupling to an additional single electron oxidation the second

NOS monooxygenation will of course lead to formation of nitroxyl, a species that is expected to undergo oxidation to NO readily under physiological conditions. Such mechanisms require further consideration if evidence of N-demethylation in the processing of NMA/NHMA is interpreted as indicating that the second NOS monooxygenation may involve oxenoid-iron rather than peroxo-iron chemistry. Moreover, recent studies by Schmidt and co-workers (*124*) suggest that NO⁻ might indeed be the immediate nitrogen oxide product of Arg-turnover by NOS.*

Variations on the mechanism proposed in Fig. 37 may also be envisaged, for example, the same oxaziridine intermediates may be generated by nucleophilic attack of PPIX–Fe(III)–OO⁻ (or PPIX–Fe(III)–OOH) on Arg and NHA. Other variable aspects of the mech-

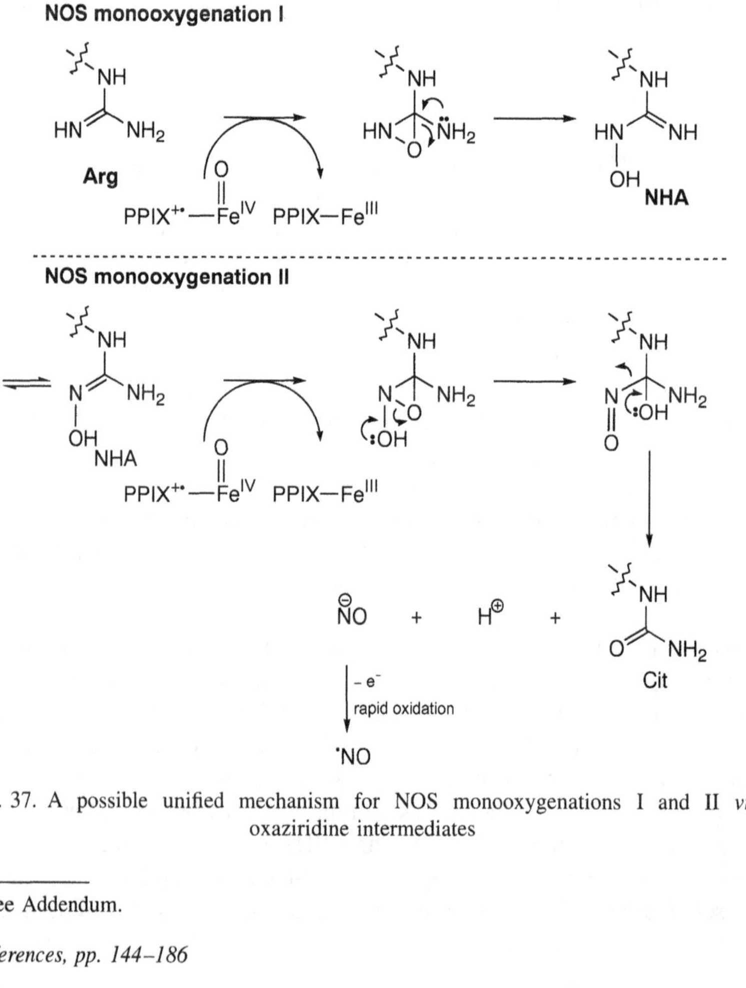

Fig. 37. A possible unified mechanism for NOS monooxygenations I and II *via* oxaziridine intermediates

---

*See Addendum.

References, pp. 144–186

anism relate to the protonation state of the substrate, the point at which the one-electron oxidation in the second reaction occurs, and whether the proposed oxaziridines would constitute distinct intermediates along the reaction pathway or reacting ensembles, depending on the timing of ring opening reactions and the degree of oxaziridine C–O bond formation.

The oxaziridination pathways could account satisfactorily for the formation of NHA from Arg and, moreover, the processing of NMA by this pathway would be expected to afford NHMA, consistent with observations discussed in section 9.4. However, this mechanism alone could not explain the demethylation of NMA to Arg and formaldehyde; alternative N-dealkylation pathways would have to operate simultaneously to account for this observation. The mechanism as formulated in Fig. 37 greatly simplifies the overall conversion of Arg into Cit and NO and, of particular importance, would presumably require minimal reorganisation of the substrate during the overall course of the two monooxygenations, since each monooxygenation would involve attack of the substrate by an identical oxoiron (or peroxide) species along a very similar trajectory. Such a simplification appears rational on evolutionary grounds and MARLETTA and co-workers (*311*) have also recently considered a similar mechanism, although it was noted that the mode of oxaziridine ring scission in the second monooxygenation would be rather unusual, and perhaps take place on the one-electron-oxidised N-hydroxyoxaziridine leading to direct formation of NO rather than NO⁻. Participation of the adjacent nitrogens as internal nucleophiles in opening the oxaziridine ring would lead to cleavage of the endocyclic C–O bond and such a mechanism would conflict with observations (*322*) that the citrulline oxygen atom is derived from O_2 and not from water. Ring opening of the oxaziridine might conceivably be assisted by a suitably positioned enzyme-bound nucleophile or organised water molecule (Fig. 38), in a manner akin to the mechanism of oxaziridine reagents used in synthetic N-aminations (*351*). This mechanism raises its own difficulties, however, in that the hypothetical enzyme-bound nucleophile would have to operate specifically in NOS monooxygenation II and not cleave the oxaziridine intermediate in the first monooxygenation.

An alternative mechanism that might assist ring cleavage of the hypothetical oxaziridine intermediates is shown in Fig. 39. In this scheme a suitably positioned basic moiety in the catalytic site deprotonates the N-hydroxy group. Such a mechanism could account for the dominance of the hydroxy oxygen over the amino nitrogens as the internal nucleophile that facilitates ring cleavage. The mechanism would also be specific to NOS monooxygenation II and not affect the corresponding oxaziridine intermediate in NOS monooxygenation I.

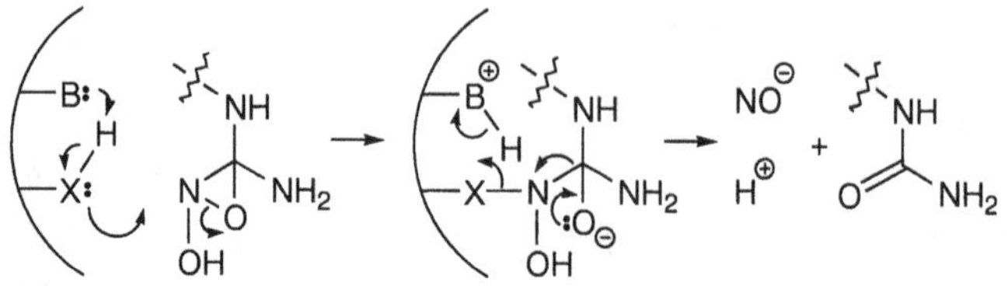

Fig. 38. A conceivable scheme for enzyme-assisted opening of an *N*-hydroxyoxaziridine intermediate. A similar mechanism can be envisaged with the substrate in a one-electron-oxidised state that might arise from ferric heme superoxide-mediated hydrogen abstraction from the substrate hydroxyl group prior to the oxaziridine forming step; in this case the reaction product would be NO rather than NO$^-$

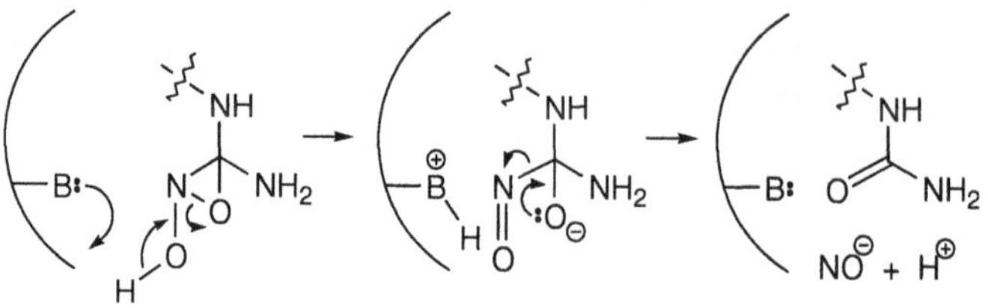

Fig. 39. An alternative scheme for enzyme-assisted opening of a hypothetical *N*-hydroxyoxaziridine intermediate in NOS monooxygenation II

In unravelling the mechanism of NOS catalysis it will also be important to account for Marletta's observations that H_2O_2, while able to support NOS monooxygenation II, is unable to facilitate the oxidation of Arg to NHA (*311*). As discussed (Section 9.3) this result could indicate that NOS-bound heme does not in fact directly mediate the first monooxygenation reaction. Biochemical characterisation of NOS monooxygenation II is possible through studies based on enzyme incubations with NHA. In contrast the separate characterisation of NOS monooxygenation I (Arg → NHA) is intrinsically difficult because of the ensuing oxygenation of NHA to Cit. Thus it is difficult to disentangle the cofactor-dependence of the first monooxygenation from that of the composite reaction from Arg to Cit. Recently, however, Giovanelli *et al.* (*335*) have found that stoichiometric amounts of purified rat brain NOS are able to convert Arg into NHA in the *absence* of added electron donor (NADPH), thus providing an experimental system that permits separate examination of the first monooxygenation. Using this system it was shown that *both* NOS monooxygenations I and II are inhibited by CO, an observation which firmly implicates heme in both steps. Moreover, in contrast to the oxidation of Arg, conversion of NHA into Cit showed an absolute requirement for addition of NADPH to the assay medium. These findings, which are summarised in Fig. 40, suggest that NOS contains an endogenous reductant, possibly H_4B, that can support a single turnover of NOS monooxygenation I but not NOS monooxygenation II. SCHMIDT and co-workers (124) have recently suggested that the number of NADPH equivalents consumed by NOS during a single catalytic cycle has been overestimated in previous studies and consequently that additional reducing equivalents from a source other than NADPH (possibly from H_4B) are necessary. It is now a matter of considerable importance that the role of H_4B in NOS catalysis be clarified.

NOS monooxygenation I: turnover of Arg
- In the absence of added NADPH: [^{3}H]-Arg $\rightarrow$ predominantly [^{3}H]-NHA; a maximum 0.16 mol of NHA is formed per mol of NOS monomer present.
- Amounts of enzyme-bound NADPH in the purified NOS used in the assay were insufficient to account for the quantity of NHA formed. (Amounts of enzyme-bound unlabelled Arg in the purified enzyme were also slight and insufficient to introduce significant error into stoichiometric calculations: no more than 0.06 mol per mol of NOS subunit.)
- Displays absolute requirement for O_2 and Ca^{++}/CaM.
- Reaction is stimulated 3-fold in presence of added H_4B.
- NHA formed is released from NOS only in small amounts though it s not as tightly bound to NOS as previously thought (326, 352).
- NHA formation shows 28% inhibition by CO.
- [^{3}H]-Arg $\rightarrow$ [^{3}H]-NHA conversion shows 98% inhibition in the presence of excess unlabelled NHA.
- Addition of NADPH causes exclusive formation of Cit instead of NHA.
- Reaction in the absence of catalase gave rise to increased amount of Cit at the expense of NHA–possibly due to the accumulation of H_2O_2 and the H_2O_2-supported conversion of NHA into Cit which has been demonstrated by MARLETTA (311).

NOS monooxygenation II: turnover of NHA
- Displays absolute requirement for NADPH.
- Displays absolute requirement for O_2 and Ca^{++}/CaM.
- Cit formation in the overall reaction shows 66% inhibition by CO.
- Reaction also stimulated by the presence of added H_4B.

Conclusions
- Heme and H_4B play a role in *both* NOS monooxygenations I and II.
- NOS contains an endogenous reductant that can support a single turnover of NOS monooxygenation I but not NOS monooxygenation II.

Fig. 40. Key findings of GIOVANELLI *et al.* (335) from incubations of [^{3}H]-Arg and [^{3}H]-NHA with purified rat brain NOS

10. Nitric Oxide Synthase Structure

10.1. Primary Structure and Domain Organisation of NOS

10.1.1. Primary Structure

NO synthases are catalytically active only in a homodimeric form, comprising, two identical protein subunits (*145, 172, 353–358*). Several reports have described the cloning and cDNA sequences for NOS isoforms from various species and cell types (*140, 142, 155–164, 180, 224, 359–372*). The primary structure for each isoform has thus been predicted and, for the three human isoforms, is shown in Fig. 41. Some 50–60% sequence identity is shared between the isoforms, with the primary structure of each individual isoform being highly conserved across different species. The primary structure of two insect NOSs have also been reported (*215, 373*). These isoforms share significant sequence identity with the mammalian enzymes, which underlines the remarkable conservation of the NOS family across vertebrates and invertebrates. The monomeric subunits of NOS possess relative molecular masses of 160–161, 133 and 130–131 kDa for the neuronal (*136, 140, 151*), endothelial (*155–158*) and inducible isoforms (*142, 145, 146, 154, 159, 160, 163, 165*) respectively. The neuronal isoform contains an N-terminal leader sequence of some 220 amino acids which is not present in the other two isoforms. This sequence does not play a role is cofactor binding, dimerisation or NO synthesis (*374*), but binds to proteins that localise the enzyme to specific intracellular sites (*375*). The other notable difference in the primary structure of the three isoforms shown is the absence in the inducible enzyme of a segment of approximately 50 amino acids that occurs in both of the constitutively expressed isoforms.

10.1.2. Domain Organisation

As shown in Fig. 41 the NADPH, FAD and FMN binding sites are all located in the C-terminal half of the enzyme which is structurally homologous to other dual-flavin enzymes such as CPR (*140*). The latter protein serves to transfer electrons from NADPH to the heme group of an associated heme protein; thus the C-terminal section of NOS was proposed to fulfil a similar reductase function. Indeed, the similarity between the C-terminal half of NOS and CPR together with the known structure of P450$_{BM-3}$ (*721*) suggests that NOS is organised into two distinct functional domains: a C-terminal reductase domain, possessing the NADPH and flavin binding sites, and an N-terminal domain in which heme and substrate binding sites are located. These two domains are linked by a short, regulatory, CaM-binding sequence.

In so far as they are known, binding sites and other significant residues are indicated on the NOS sequences shown in Fig. 41. The location of binding sites within the reductase domain is straightforward because of homology to CPR. Mutational analysis, limited proteolysis and expression of truncated forms of the NOS oxygenase domain have allowed identification of the various key residues in the N-terminal half of the NOS sequences.

The bidomain structure of NOS has been confirmed by limited trypsin proteolysis which cleaves NOS between its two functional domains. In the case of nNOS trypsin-sensitive regions are located at the junctions of the oxygenase domain with the N-terminal leader and the CaM-binding region (*390, 391*). A related study with iNOS, which is primarily cleaved into 2 fragments by a cut at the oxygenase/reductase junction, confirmed a similar structural organisation for this isoform and additionally localised the H_4B binding site to the N-terminal oxygenase domain of the enzyme (*392*). The individual reductase and oxygenase domains of NOS have also been separately expressed in bacteria, yeast and insect cells (*385, 386, 393–395*). Although the isolated NOS fragments produced by trypsinolysis or individual expression lack the capacity for NO synthesis, the cofactor content and spectral properties of the two domains are unaltered. Moreover, the functional properties of the reductase domain are retained, in that the isolated domain catalyses electron transfer from NADPH to acceptors such as cytochrome *c* or ferricyanide at rates comparable to those observed with full-length NOS proteins (*393, 394*). The isolated oxygenase domain is unable to catalyse NO formation in a medium replete with NADPH, H_4B, O_2 and Arg. However, in the presence of H_2O_2 NHA is converted into NO and Cit (*396*), a result that accords with MARLETTA'S studies (*311*) with the full-length enzyme (Section 9.3). Retention of the NOS monooxygenation II capacity suggests that the functional properties of the isolated oxygenase domain, like the reductase portion of the enzyme, are maintained. In summary, the separated oxygenase and reductase domains of NOS retain their functional activity and thus appear to be able to fold, exist and function independently of one another (*391*); studies have shown that NOS activity can be successfully reconstituted by mixing the isolated domains, although catalytic activity is suboptimal in comparison to the full-length enzymes (*394, 397*).

```
MEDHMFGVQQIQPNVISVRLFKRKVGGLGFLVKERVSKPPVIISDLIRGGAAEQSGLIQAGDIILAVNGRPLVDLSYDSALEVLRGIASETHVVLILRGP   0100   nNOS
...................................................................................................

EGFTTHLETTFTGDGTPKTIRVTQPLGPPTKAVDLSHQPPAGKEQPLAVDGASGPGNGPQHAYDDGQEAGSLPHANGLAPRPPGQDPAKKATRVSLQGRG   0200   nNOS
...................................................................................................

                         1             2            3
ENNELLKEIEPVLSLLTSGSRGVKGGAPAKAEMKDMGIQVDRDLDGKSHKPLPLGVENDRVFNDLWGKGNVPVVLNNPYSEKEQPPTSGKQSPTKNGSPS   0300   nNOS
.....................MGNLKSVAQEPGPPCGLGLGLGLGLCGKQGPATPAPEPSRAPASLLPPAPEHSPPSSPLTQPP                    0063   eNOS
................MACPWKFLFKTKFHQYAMNGEKDINNNVEKAPCATSSPVTQDDLQYHNLSKQQNESPQPLVETGKKSPESLVKLDATPL         0079   iNOS

                   4                                  - -5- -
KCPRFLKVKNWETEVVLTDTLHLKSTLETGCTEYICMGSIMHPSQHARRPEDVRTK·GQLFPLAKEFIDQYYSSIKRFGSKAHMERLEEVNKEIDTTSTY   0399   nNOS
EGPKFPRVKNWEVGSITYDTLSAQAQQDGPCTPRRCLGSLVFPRKLQGRPSPGPPAPEQLLSQARDFINQYYSSIKRSGSQAHEQRLQEVEAEVAATGTY   0163   eNOS
SSPRHVRIKNWGSGMTFQDTLHHKAKGILTCRSKSCLGSIMTPKSLTRGPRDKPTPPDELLPQAIEFVNQYYGSFKEAKIEEHLARVEAVTKEIETTGTY   0179   iNOS

            6
QLKDTELIYGAKHAWRNASRCVGRIQWSKLQVFDARDCTTAHGMFNYICNHVKYATNKGNLRSAITIFPQRTDGKHDFRVWNSQLIRYAGYKQPDGSTLG   0499   nNOS
QLRESELVFGAKQAWRNAPRCVGRIQWGKLQVFDARDCRSAQEMFTYICNHIKYATNRGNLRSAITVFPQRCPGRGDFRIWNSQLVRYAGYRQQDGSVRG   0263   eNOS
QLTGDELIFATKQAWRNAPRCIGRIQWSNLQVFDARSCSTAREMFEHICRHVRYSTNNGNIRSAITVFPQRSDGKHDFRVWNAQLIRYAGYQMPDGSIRG   0279   iNOS

                                                  - - - - - - - -7- - - - - - - - -        8
DPANVQFTEICIQQGWKPPRGRFDVLPLLLQANGNDPELFQIPPELVLEVPIRHPKFEWFKDLGLKWYGLPAVSNMLLEIGGLEFSACPFSGWYMGTEIG   0599   nNOS
DPANVEITELCIQHGWTPGNGRFDVLPLLLQAPDEPPELFLLPPELVLEVPLEHPTLEWFAALGLRWYALPAVSNMLLEIGGLEFPAAPFSGWYMSTEIG   0363   eNOS
DPANVEFTQLCIDLGWKPKYGRFDVVPLVLQANGRDPELFEIPPDLVLEVAMEHPKYEWFRELELKWYALPAVANMLLEVGGLEFPGCPFNGWYMGTEIG   0379   iNOS

          9                                            9  9                       9
VRDYCDNSRYNILEEVAKKMNLDMRKTSSLWKDQALVEINIAVLYSFQSDKVTIVDHHSATESFIKHMENEYRCRGGCPADWVWIVPPMSGSITPVFHQE   0699   nNOS
TRNLCDPHRYNILEDVAVCMDLDTRTTSSLWKDKAAVEINVAVLHSYQLAKVTIVDHHAATASFMKHLENEQKARGGCPADWAWIVPPISGSLTPVFHQE   0463   eNOS
VRDFCDVQRYNILEEVGRRMGLETHKLASLWKDQAVVEINIAVLHSFQKQNVTIMDHHSAAESFMKYMQNEYRSRGGCPADWIWLVPPMSGSITPVFHQE   0479   iNOS
                                                                            GDSHVDTSSTVSEAVAEEV         0019   CPR

                     - - - - - - -Ca²⁺/CaM- - - - - -
MLNYRLTPSFEYQPDPWNTHVWKGTNGTPTKRRAIGFKKLAEAVKFSAKLMGQAMAKRVKATILYATETGKSQAYAKTLCEIFKHAFDAKVMSMEEYDIV   0799   nNOS
MVNYFLSPAFRYQPDPWKGSAAKGT·GI·TRKK··TFKEVANAVKISASLMGTVMAKRVKATILYGSETGRAQSYAQQLGRLFRKAFDPRVLCMDEYDVV   0559   eNOS
MLNYVLSPFYYYQVEAWKTHVWQDEKRRP·KRREIPLKVLVKAVLFACMLMRKTMASRVRVTILFATETGKSEALAWDLGALFSCAFNPKVVCMDKYRLS   0578   iNOS
SLFSMTDMILFSLIVGLLTYWFLFRKKKEEVPEFTKIQTLTSSVRESSFVEKMKKTGRNIIVFYGSQTGTAEEFANRLSKDAHRYGMRGMSADPEEYDLA   0119   CPR

                                                                                    - - -FMN- - - -
HLEHETLVLVVTSTFGNGDPPENGEKFGCALMEMRHP·NS··VQEERKSYKVRFNSVSSYSDSQKSSGD·GPDLRDNFESAGPLANVRFSVFGLGSRAYP   0895   nNOS
SLEHETLVLVVTSTFGNGDPPENGESFAAALMEMSGPYNSSPRPEQHKSYKIRFNSISC·SDPLVSSWRRKRKESSNTDSAGALGTLRFCVFGLGSRAYP   0658   eNOS
CLEEERLLLVVTSTFGNGDCPGNGEKLKKSLFMLKELNN·······································KFRYAVFGLGSSMYP           0632   iNOS
DLSSLPEIDNALVVFCMATYGEGDPTDNAQDFYDWLQETDVD·····································LSGVKFAVFGLGNKTYE        0178   CPR

- - - - - -FMN- - - - - - - - - - - - -
HFCAFGHAVDTLLEELGGERILKMREGDELCGQEEAFRTWAKKVFKAACDVFCVGDDVNIEKANNSLISNDRSWKRNKFRLTFVAEAPELTQGLSNVHKK   0995   nNOS
HFCAFARAVDTRLEELGGERLLQLGQGDELCGQEEAFRGWAQAAFQAACETFCVGEDA··KAAARDIFSPKRSWKRQRYRLSAQAEGLQLLPGLIHVHRR   0756   eNOS
RFCAFAHDIDQKLSHLGASQLTPMGEGDELSGQEDAFRSWAVQTFKAACETFDVRGKQ··HIQIPKLYTSNVTWDPHHYRLVQDSQPLDLSKALSSMHAK   0730   iNOS
HFNAMGKYVDKRLEQLGAQRIFELGLGDDDGNLEEDFITWREQFWPAVCEHFGVEATGEESSIRQYELVVHTDIDAAKVYMGEMGRLKSYENQKPPFDAK   0278   CPR
```

```
                   FAD pyrophosphate
RVSAARLLSRQNLQSPKSSRSTIFVRLHTNGSQELQYOPGDHLGVFPGNHEDLVNALIERLEDAPPVNQMVKVELLEERNTALGVISNWTDELRLPPCTI   1095   nNOS
KMFQATIRSVENLQSSKSTRATILVRLDTGGQEGLQYOPGDHIGVCPPNRPGLVEALLSRVEDPPAPTEPVAVEQL·EKGSPGGPPPGWVRDPRLPPCTL   0855   eNOS
NVFTMRLKSRQNLQSPTSSRATILVELSCEDGQGLNYLPGEHLGVCPGNQPALVQGILERVVDGPTPHQTVRLEALDESGS······YWVSDKRLPPCSL   0824   iNOS
NPFLAAVTTNRKLNQG·TERHLMHLELDISDSK·IRYESGDHVAVYPANDSALVNQLGKIL··GADLDVVMSLNNLDEESNKKHPFPC········PTSY    0366   CPR

       electron transfer                                                              FAD isoalloxazine
EQAFKYYLDITTPPTPLQLQQFASLATSEKEKQRLLVLSK····GLQEYEEWKWGKNPTIVEVLEEFPSIQMPATLLLTQLSLLQPRYYSISSSPDMYPD   1191   nNOS
RQALTFFLDITSPPSPQLLRLLSTLAEEPREQQELEALSQ····DPRRYEEWKWFRCPTLLEVLEQFPSVALPAPLLLTQLPLLQPRYYSVSSAPSTHPG   0951   eNOS
SQALTYFLDITTPPTQLLLQKLAQVATEEPERQRLEALCQ·····PSEYSKWKFTNSPTFLEVLEEFPSLRVSAGFLLSQLPILKPRFYSISSSRDHTPT   0919   iNOS
RTALTYYLDITNPPRTNVLYELAQYASEPSEQELLRKMASSSGEGKELYLSWVVEARRHILAILQDCPSLRPPIDHLCELLPRLQARYYSIASSSKVHPN   0466   CPR
                                                                             *         *  **  *

                                                  ----NADPH ribose---
EVHLTVAIVSYRTRDGEGPIHHGVCSSWLNRIQA······DELVPCFVRGAPSFHLPRNPQVPCILVGPGTGIAPFRSFWQORQFDIQHKGMNPCPMVLV   1285   nNOS
EIHLTVAVLAYRTQDGLGPLHYGVCSTWLSQLKP······GDPVPCFIRGAPSFRLPPDPSLPCILVGPGTGIAPFRGFWQERLHDIESKGLQPTPMTLV   1045   eNOS
EIHLTVAVVTYHTRDGQGPLHHGVCSTWLNSLKP······QDPVPCFVRNASGFHLPEDPSHPCILIGPGTGIAPFRSFWQORLHDSQHKGVRGGRMTLV   1013   iNOS
SVHICAVVVEYETKAGR··INKGVATNWLRAKEPAGENGGRALVPMFVRK·SQFRLPFKATTPVIMVGPGTGVAPFIGFIQERAW·LRQQGKEVGETLLY   0562   CPR
          **                                                           ***  *

                                               --NADPH adenine--
     †                                                     †
FGCRQSKIDHIYREETLQAKNKGVFRELYTAYSREPDKPKKYVQDILQEQLAESVYRALKEQGGHIYVCGDVT·MAADVLKAIQRIMTQQGKLSAEDAGV   1384   nNOS
FGCRCSQLDHLYRDEVQNAQQRGVFGRVLTAFSREPDNPKTYVQDILRTELAAEVHRVLCLERGHMFVCGDVT·MATNVLQTVQRILATEGDMELDEAGD   1144   eNOS
FGCRRPDEDHIYQEEMLEMAQKGVLHAVHTAYSRLPGKPKVYVQDILRQQLASEVLFVLHKEPGHLYVCGDVR·MARDVAHTLKQLVAAKLKLNEEQVED   1112   iNOS
YGCRRSDEDYLYREELAQFHRDGALTQLNVAFSRE·QSHKVYVQHLLKQD·REHLWKLI·EGGAHIYVCGDARNMARDVQNTFYDIVAELGAMEHAQAVD   0659   CPR
          **       * * *                                              **

         -NADPH-
FISRMRDDNRYHEDIFGVTLRTYEVTNRLRSESIAFIEESKKDTDEVFSS                                                    1434   nNOS
VIGVLRDQQRYHEDIFGLTLRTQEVTSRIRTQSFSLQERQLRGAVPWAFDPPGSDTNSP                                            1203   eNOS
YFFQLKSQKRYHEDIFGAVFPYEAKKDRVAVQPSSLEMSAL                                                             1153   iNOS
YIKKLMTKGRYSLDVWS                                                                                     0676   CPR
           *
```

Fig. 41. Primary structure and sequence analysis for the 3 human NOS isoforms and CPR. Consensus binding sites for FMN, FAD and NADPH, assigned by analogy to CPR (*377*) and ferredoxin-nicotinamide adenine phosphate reductase (*378*), are indicated as is the CaM binding sequence. Other marked residues are as follows: **1**, eNOS myristoylation site (*156–158, 379*) **2** and **3**, eNOS Cys residues that are subject to palmitoylation (*380*); **4**, Cys residue required for high-affinity binding of H_4B (*381, 382*); **5**, cAMP-dependent protein kinase phosphorylation site; **6**, Cys residue that contributes proximal axial thiolate ligand for heme (C420, C184 and C200 in human nNOS and iNOS respectively (*381, 383–386*); corresponds to C415 in rat nNOS and C194 murine iNOS); **7**, sequence that is critical to interdomain electron transfer of electrons onto the heme (*387*); **8**, glutamic acid residue that binds substrate's guanidine (*298, 299*); **9**, residues required for subunit dimerisation (*299, 388*); *, conserved residues proposed as contacts with FAD and NADPH (*215*); †, conserved Cys residues critical for NADPH binding and/or electron transfer from NADPH to FAD (*389*)

10.2. Function of the NOS Reductase Domain

By analogy with CPR it is clear that the function of the NOS reductase domain is to serve both as a conduit and as a reservoir of electrons required for reductive activation of O_2 and substrate oxidation. MARLETTA *et al.* (*384*) have reported the expression of full length C415H, C415A and C415S mutants of rat nNOS, which lack the heme prosthetic group and have been used to characterise the reductase domain of NOS. The UV-visible spectra of the mutants are essentially identical and contain features of an oxidised flavoprotein similar to CPR. Quantitative oxidative titration with ferricyanide suggests that the isolated enzymes exist in a one-electron-reduced state possessing one flavin in the fully oxidised form and the other in the neutral semiquinone form. The flavin semiquinone absorbance has also been observed in wild-type nNOS (*166, 383, 398*). In an important development McMILLAN and MASTERS have expressed the N-terminal oxygenase domain (residues 1–714) and C-terminal reductase domain (residues 715–1429) of nNOS as separate proteins (*385*). The isolated functional domains were found to retain properties of the intact enzyme, with the reductase domain retaining a UV-visible spectrum characteristic of an oxidised flavoprotein and, on reduction with NADPH, giving rise to an air-stable semiquinone similar to the one-electron-reduced FAD-FMNH˙ form of CPR (*399–401*). In contrast, the spectrum of apo-NOS, a heme-deficient enzyme produced by prolonged dialysis with urea, lacks the flavin semiquinone absorbance; this discrepancy is possibly due to its destabilisation resulting from protein disruption (*402*).

In the case of CPR, anaerobic reduction of the oxidised protein with 2 equivalents of NADPH affords the three-electron-reduced enzyme containing a FADH˙–$FMNH_2$ flavin combination (*400*). The inability of stoichiometric amounts of NADPH to reduce CPR exhaustively and the relative stability of the one-electron-reduced form to oxidation suggests that CPR shuttles between the one-and three-electron-reduced states, though exhaustive reduction and oxidation of CPR can be mediated chemically by dithionite and ferricyanide respectively (Fig. 42). The fully reduced flavin, $FMNH_2$, transfers electrons from CPR to the cytochrome P450 heme. MARLETTA *et al.* have proposed (*384*) that a similar shuttling of flavin redox states occurs in the NO synthases; the electron transfer pathway in NOS is summarised in Fig. 43. A full redox cycle model has been developed by GRIFFITH and STUEHR (*119*) to account for the overall NADPH stoichiometry of the Arg–NO pathway, which, according to the prevailing view, assumes that all reducing

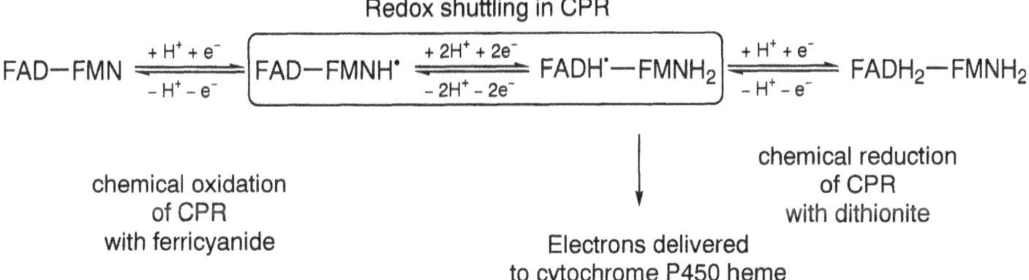

Fig. 42. Electron cycling between flavin cofactors of CPR

equivalents are supplied by NADPH and that the product of NOS turnover is NO rather than NO^-.

EPR studies (*309, 398*) with the neuronal and endothelial isoforms of NOS have confirmed the presence of an $FMNH^{\cdot}$ flavin semiquinone in the resting enzymes; this cofactor gives rise to a prominent radical signal in the EPR spectra of the enzymes. STUEHR and co-workers (*398*) have observed spin-spin coupling between the flavin and heme groups, thereby showing that these two redox centres must be positioned close to one another, consistent with the requirement for interdomain electron transfer. Dissolved O_2 was found to affect the $FMNH^{\cdot}$ radical and this was taken as an indication that the FMN-binding site may be partially exposed to solvent. However, transfer of electrons from the isolated NOS reductase domain to O_2 is very slow (*172, 392–394*). Thus the reductase

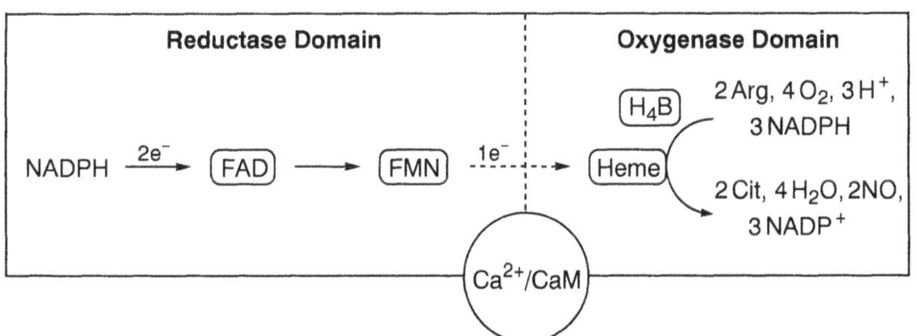

Fig. 43. Schematic representation of electron flow and catalytic function of NOS. NADPH, an obligate two electron donor, supplies electrons to the reductase domain. The flavins of the reductase domain have a capacity for holding upto 4 electrons in principle, but in practice jointly hold between one and three at any given point. These electrons are delivered singly as required for reductive activation of O_2 at the heme centre and subsequent substrate oxidation. Electron flux between reductase and oxygenase domains requires the presence of bound CaM

domain itself is a poor generator of superoxide even though full-length NOS generates superoxide under Arg-deficient conditions. This is consistent with a functional model for NOS in which electron transfer from FMN to the heme is followed by heme-catalysed electron transfer to O_2.

In studies of eNOS, Tsai et al. (*309*) have found the spin relaxation properties of the FMNH· radical to indicate the presence of a dipolar interaction between the heme and flavin centres. The amount of EPR-detectable radical exhibited no correlation with the concentration of the high-spin heme. Indeed, the spectral characteristics of the flavin radical remained the same regardless of whether the eNOS heme state was exclusively low-spin (as the imidazole complex) or exclusively high-spin (as the Arg complex). These analyses led to the conclusion that it is most likely the heme redox potential rather than spin state that dictates the amount of radical formed. Moreover, the addition of Ca^{2+} and CaM to the eNOS samples had little effect on the properties of the flavin radical. This is an important observation because a small change in the physical distance between the flavin centre in the reductase domain and the heme centre in the oxygenase domain is expected to exhibit a pronounced effect on the EPR properties of the flavin radical because of the influence of the dipolar interaction on its relaxation. These results do not support hypotheses previously advanced (*119, 183*) that Ca^{2+}/CaM binding to NOS triggers electron transfer between the domains by bringing about a significant change in the separation between the FMN and heme centres. Instead the functional role of Ca^{2+}/CaM binding may be to facilitate electron flux by reorientation of the heme and flavin centres with respect to one another or by altering the redox potential of the flavin and/or heme.

There is some debate at present over the precise stoichiometry of NADPH consumed in the Arg-NO pathway. An overall NADPH:Arg stoichiometry of 1.5:1 is required for generation of NO. NADPH stoichiometries higher than 1.5 could indicate that NO^- rather than NO is formed by NOS and this possibility has been considered (*403*). Unfortunately, determination of NADPH stoichiometry is somewhat complicated by the uncoupling* of NADPH-dependent oxygen activation from substrate oxidation in a proportion of catalytic cycles (*167*).

---

*Under maximal conditions electron transfer in NOS is tightly coupled to O_2 and to substrate oxidation. This is not the case, however, when NOS is incubated under Arg-deficient conditions (*183, 345, 404*) and during the processing of certain NOS inhibitors such as NMA (*343*).

When that happens reactive oxygen species (ROS), superoxide and H_2O_2, are formed at the expense of NADPH consumption (*345, 404*).

Recently SCHMIDT and co-workers have concluded (*124*) that NADPH:Arg stoichiometry has been overestimated because of the consumption of NADPH by peroxynitrite, the latter being formed by the reaction of NO with superoxide present in the assay medium. Superoxide is invariably present under assay conditions, formed either by uncoupling of the NOS reaction or by autoxidation of H_4B, but can be suppressed by the addition of SOD. These authors found the presence of SOD to lower the NADPH stoichiometry (measured with respect to Cit formed) to approximately 1:1 however. Moreover, additional observations by the group indirectly supported the idea that NO is not formed as the immediate product of NOS-catalysed Arg turnover, but by subsequent oxidation of a precursor molecule such as nitroxyl.* Additional reducing equivalents are required from a source other than NADPH if the NADPH:Arg stoichiometry is indeed lower than previously thought and at same time the proximal product of NOS catalysis is not NO but nitroxyl. SCHMIDT *et al.* tentatively proposed a redox role for H_4B to account for the additional reducing equivalents required by this scenario.

GIOVANELLI and co-workers have found that substoichiometric amounts of NHA are produced from Arg by partial NOS turnover *in the absence of NADPH* (*335*). These authors have, therefore also raised the question of whether H_4B is capable of supporting substrate oxidation in NOS monooygenation I to account for their observations. Taken together, the results of SCHMIDT and GIOVANELLI show that it will be essential to define the redox role of H_4B and unequivocally establish the identity of the proximal product of NOS catalysis in order to progress in our understanding of the mechanism that operates in these complex enzymes.

10.3. The Calmodulin Binding Site

CaM binding to NOS triggers the transfer of NADPH-derived electrons from the reductase domain onto the heme and is essential for NO synthesis. STUEHR and ABU-SOUD found that CaM binding to the neuronal isoform in response to elevated Ca^{2+} levels caused rapid oxidation of NADPH and was independent of substrate binding to the

---

* See Addendum.

enzyme (*183*). In contrast, the inducible isoform was found to contain tightly bound CaM and, therefore, its activity is independent of increases in Ca^{2+} levels.

Further work by Stuehr and associates points to a dual mechanism of NOS activation by CaM (*402*). This discovery was made by investigating the ability of NOS to reduce external oxidants such as cytochrome c in a manner analogous to CPR, which is homologous to the NOS reductase domain. In these studies Stuehr *et al.* found that apo-NOS, an enzyme lacking bound heme and H_4B, was also capable of reducing cytochrome c, a result which indicated that the major electron pathway for reduction of external oxidants does not involve the NOS heme. The binding of Ca^{2+} / CaM to NOS was found to increase the rate of NADPH oxidation by cytochrome c, thus revealing an *intra*domain acceleration of electron transfer. Similar behaviour has also been observed (*299, 394*) for truncated NOS proteins consisting of the reductase domain alone. In these truncated proteins CaM binding stimulates electron transfer to cytochrome c by the same magnitude observed in the full-length enzyme (*393*). This analysis has been supported by further with studies of nNOS mutants that lack heme and H_4B cofactors (*384*). CaM binding must, therefore, elicit redox and / or conformational changes in the NOS reductase domain itself which accelerate the rate by which NADPH reduces the flavins. Additionally the binding of CaM to NOS may regulate enzyme activity by modulating interdomain electron transfer. A number of other groups have characterised NO synthases by their ability to reduce cytochrome c (*172, 391, 405*).

Although CaM binding to NOS influences electron flux in the enzyme, it does not affect the binding of ligands such as cyanide or CO to the heme iron nor does it alter the NOS binding affinity for Arg (*406*). Therefore CaM does not regulate NOS activity by controlling Arg binding to the enzyme. Apart from sensitivity to levels of Ca^{2+} ions, the modulation of NOS activity by CaM may also involve a more complex interaction with other cellular factors. For example, physiological concentrations of the pineal hormone melatonin have been shown (*407*) to inhibit nNOS by interaction with CaM, potentially by modification of its binding to the CaM recognition site of the enzyme.

CaM is responsible for the activation of numerous enzymes when the intracellular Ca^{2+} ion concentration is elevated from resting levels ($< 10^{-7}$ M) to the micromolar range [see Persechini (*408*) and references therein]. The protein is dumbbell-shaped and comprises two globular lobes linked by a shared central helix. Binding of two Ca^{2+} ions to each of the CaM domains induces a conformational change in the protein that exposes two methionine rich hydrophobic surfaces. These surfaces facilitate interaction with the regulated protein, which usually possesses an uninterrupted sequence of about 20 amino acids as its CaM-binding domain [see ZHANG (*409*) and references therein]. A number of studies have begun to throw light on the nature of the binding interaction between NOS and CaM. In particular useful work (*410–415*) has been carried out with synthetic 28–35-residue peptides that correspond to the NOS CaM binding sites and bind CaM with an affinity similar to their parent enzymes. The results of these studies suggest that CaM binding is indeed determined primarily by a small segment of the NOS protein and that sequence differences within the CaM-binding region confer different affinities for CaM between the isoforms. In the case of rat nNOS a specific hydrophobic / basic amino acid cluster within the CaM-binding domain ($F_{731}K_{732}K_{733}L_{734}$) that is critical for binding of CaM has been identified (*416*). Specific amino acids ($F_{498}K_{499}$ and L_{511}) critical for the interaction with CaM have also been disclosed in the CaM-binding sequence of bovine eNOS (*414*).

VOGEL and co-workers (*490, 417*) have examined the interaction between CaM and a 23-residue synthetic peptide encompassing the CaM-binding domain of rat cerebellar NOS (residues 725–747: KRRAIGFKKLAEAVKFSAKLMGQ). The nNOS peptide and CaM were found to bind in a head-to-tail arrangement with a mode of binding typical of other CaM-protein complexes. PERSECHINI *et al.* have also studied the interaction between CaM and NOS (*408, 411, 418, 419*) and suggest a model for CaM-dependent activation of the enzyme in which the NOS CaM-binding domain functions as an intrasteric inhibitor, the influence of which is relieved when bound by CaM. Ca^{2+} ion dissociation from the N-terminal lobe of CaM is thought to be coupled to NOS inactivation and a slower Ca^{2+} ion dissociation from the C-terminal CaM lobe is linked to dissociation of the CaM–NOS complex (*411*).

In contrast to eNOS and nNOS the inducible isoform is a Ca^{2+}-independent enzyme. In this case CaM remains tightly bound to the enzyme even in the absence of elevated intracellular Ca^{2+} levels (*196*). KOCH *et al.* (*415*) have recently compared the properties of the CaM-binding sections of rat nNOS and murine iNOS and suggested that the extremely tight binding of CaM to iNOS is mediated solely by interaction of this segment of the protein and not by other parts. However, studies with chimeric NOS enzymes (*414, 420, 421*) have led to speculation that other regions of the NOS protein may play a role in determining the binding affinity of the enzyme for CaM. Recent work from STUEHR's laboratory (*422*) has found murine iNOS expressed in bacteria in the absence of CaM to be monomeric, devoid of flavins and heme, and to lack NO synthesis activity. This suggests that iNOS is improperly folded when expressed under CaM-free conditions and that tight binding of CaM to its 28 residue recognition site on the enzyme is independent of the rest of the protein. Therefore, CaM binding to iNOS may occur immediately after translation and play an active role in folding and stabilising the enzyme. The constitutive isoforms, on the other hand, exist in their CaM-free form in cells and, even when expressed in bacteria in the absence of CaM (*385, 423*), fold correctly and bind other cofactors.

10.4. Membrane Association of NOS

The neuronal and inducible NOS isoforms have generally been regarded as predominantly cytostolic and have been found in the soluble fraction of cell and tissue homogenates (*119, 143, 197, 235, 323*). In contrast eNOS is a particulate enzyme which partitions with the membrane fraction of endothelial cells (*424*). Some reports, however, suggest that a significant fraction of both nNOS (*248, 425–428*) and iNOS (*186, 429*) may also be localised to membranes. Thus, recent work reveals that a substantial fraction of nNOS is associated with endo-plasmic reticulum and specialised electron-dense postsynaptic mem-brane regions in neurons (*425, 427, 430*), and with the sarcolemma* in skeletal muscle (*250*). Moreover, other work has demonstrated that eNOS can undergo translocation from the membrane to the cytosol (*192*). The relationship between NOS function and membrane associa-tion is, therefore, a complex one which several groups are attempting to unravel at present.

Sequence alignment of the three mammalian NOS isoforms reveals the presence of an N-terminal leader sequence in nNOS (*ca.* 220 amino acids) which is absent from iNOS and eNOS. This leader sequence arises from the occurrence of three exons in the nNOS gene (*181*) which are lacking in the iNOS and eNOS genes (*163, 180*). Contained within the nNOS leader is a PDZ domain (*248, 375, 432*) that extends between

---

* The sarcolemma is the electrically excitable cell membrane of striated muscle fibers (*i.e.* skeletal muscle cells). The membrane has a complex structure and is reinforced by an actin-containing cytoskeleton which is linked to the extracellular matrix by a complex of intracellular and transmembrane proteins (Fig. 45); this network is formed around dystrophin and related proteins (*431*).

References, pp. 144–186

residues 16 and 130 (*433*). PDZ domains* have recently emerged as a class of protein-recognition modules which facilitate protein-protein binding, important, for example, to the clustering of membrane proteins within specialised multiprotein complexes that participate in signalling pathways. For example, PDZ domains of PSD-95, a brain synaptic protein containing three such modules in tandem, facilitate binding to protein subunits of certain K^+ channels and *N*-methyl-D-aspartic acid (NMDA) receptor ion channels; protein clustering is mediated by interaction between the PDZ domains of PSD-95 and specific C-terminal motifs of 3–7 residues in the ion channel and receptor proteins (*436–439*). Within signalling proteins, then, PDZ domains serve to pinpoint their activity to specific submembranous multiprotein complexes. The crystal structures of two PDZ domains have recently been reported (*440, 441*). The domains comprise a sequence of typically 80–100 amino acids which contain a GLGF signature and form a carboxylate-binding loop structure to recognise the C-terminus of a protein ligand (Fig. 44).

Although PDZ domains in different proteins vary in their ligand binding specificity, any given PDZ domain may be able to complex several binding partners which share a common C-terminal sequence. In a recent study by Li *et al.* (*433*) 13 billion peptides were screened to identify potential ligands for the PDZ domain of nNOS. Peptides ending in a DXV motif at the C-terminus were found to bind tightly to the enzyme. The PDZ domains of PSD-95, by contrast, bind peptides ending in (S/T)XV, with specificity for S/T at the -2 position arising from an interaction with $H_{372}E_{373}$ residues in PSD-95. The corresponding residues in the nNOS PDZ domain are $Y_{77}D_{78}$ (Fig. 44) and, predictably, mutation of these residues to $H_{77}E_{78}$ changed the nNOS PDZ domain binding specificity from DXV to TXV. Using the terminal DXV consensus sequence Li and co-workers also conducted a protein database search for potential nNOS binding partners. Glutamate and melatonin receptors were among the proteins identified and are of significance because of their possible involvement in NO signalling. Thus, glutamate (*246*) and melatonin (*442*) may stimulate NO biosynthesis, although

---

* PDZ domains, also known as GLGF repeats or discs-large homologous regions (DHRs), are peptidic modules that occur within a diverse set of membrane-bound proteins associated with cell junctions such as synapses of the central nervous system (CNS). The PDZ domain is named (*434*) after three proteins in which it has been identified: a 95 kDa postsynaptic density protein (PSD-95 also called SAP90), the *Drosophila* septate junction protein discs-large (Dlg) and the epithelial tight-junction protein zona occludens-1 (ZO-1). Many other proteins contain PDZ domains including nNOS (*181, 432*). For recent reviews of the role of PDZ domains in clustering signalling proteins see BREDT (*435*) and PONTING (*436*).

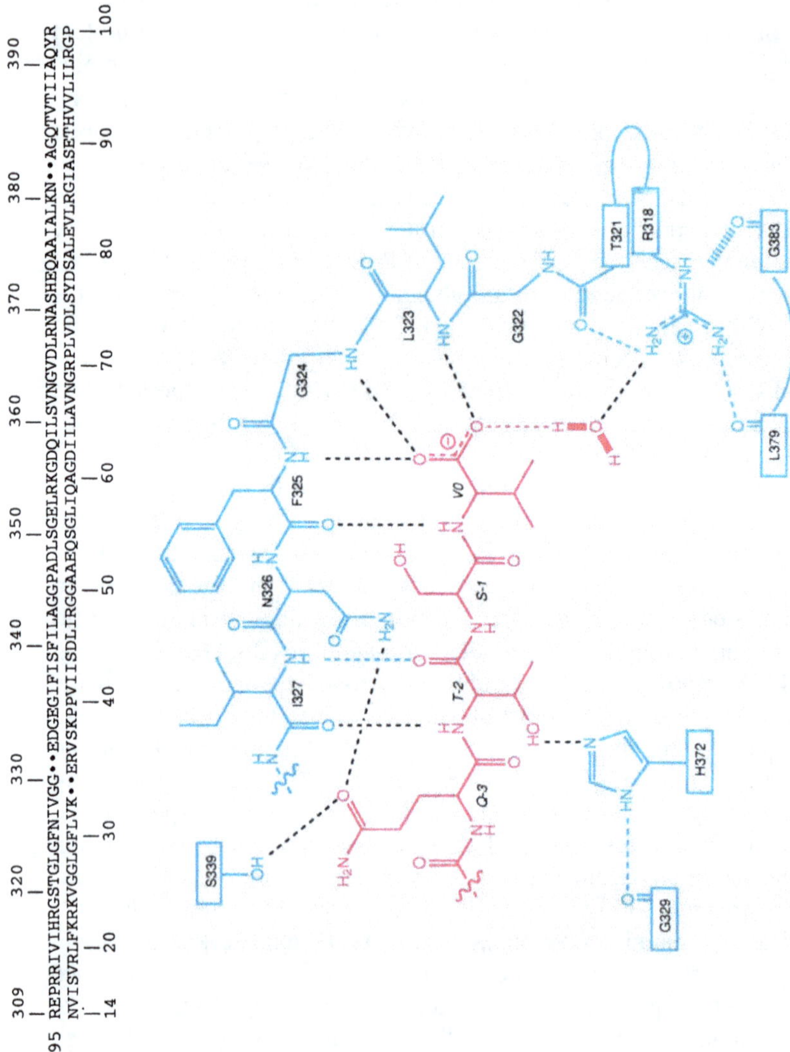

Fig. 44 (Legend on page 86)

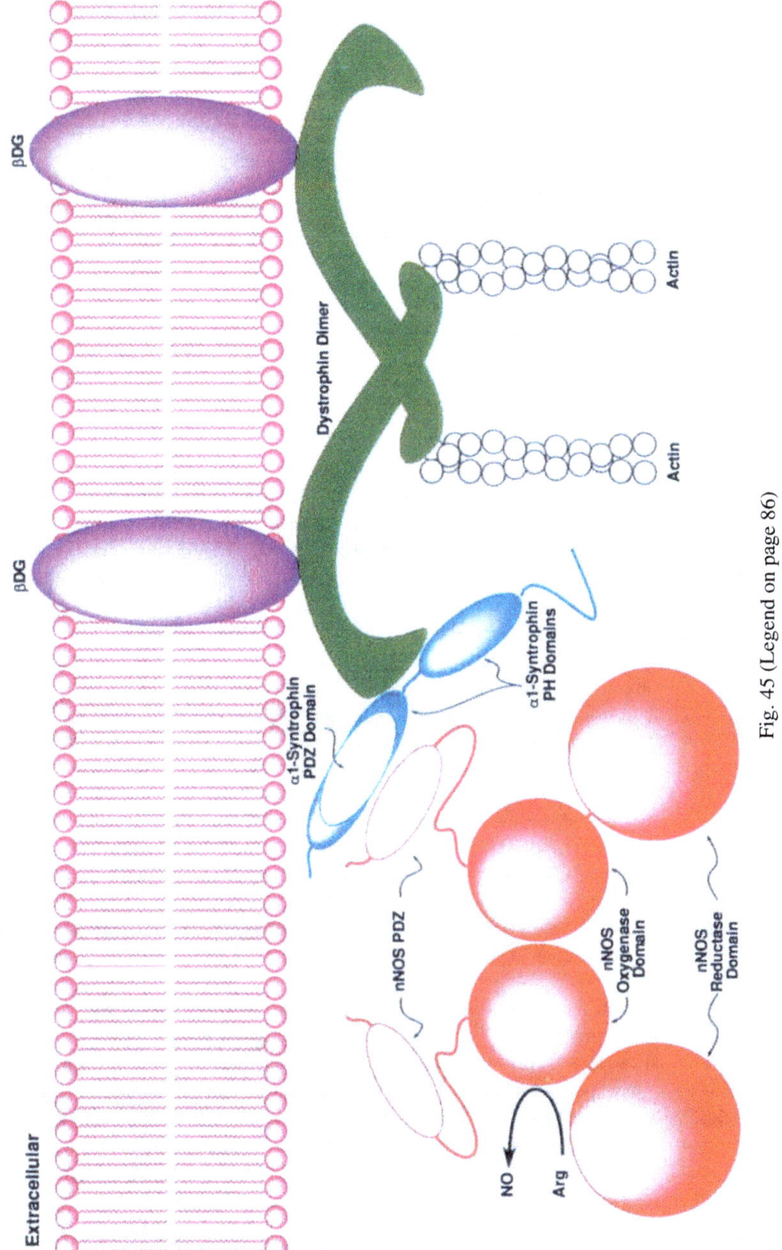

Fig. 45 (Legend on page 86)

calcium influx through NMDA receptors is the major factor that activates nNOS. The involvement of melatonin in NO signalling, however, may be more complex as a recent report (*407*) has described the inhibition of nNOS by this hormone, which apparently interacts with CaM and modifies binding of the latter to the enzyme. Li's study (*433*) has recently been corroborated by the results of another search for potential ligands of the nNOS PDZ domain undertaken by Hendriks and co-workers (*443*). This work identified the C-terminal consensus sequence for preferred recognition by nNOS as G(D/E)XV and confirmed that the enzyme's PDZ domain can mediate binding to several other proteins in the brain; potential nNOS targets included a melanoma-associated antigen, cyclophilins and the α1C-adrenergic receptor.

Fig. 44. Sequence alignment of the nNOS PDZ domain with the third PDZ domain (PDZ3) of the synaptic protein PSD-95. The crystal structure of PDZ3 complexed with a pentapeptide ligand (KQTSV) has been reported (*440*). The domain has a compact globular structure of 25–30 Å diameter. Complexation, facilitated by an exposed groove on the protein surface ending in a carboxylate binding GLGF loop, is shown schematically: heavy lines represent the peptidic backbone of PDZ3; dashed lines represent hydrogen bonds between the ligand and PDZ3. The ligand engages PDZ3 through antiparallel main chain interactions with a β sheet of PDZ3, through interactions of its terminal carboxylate with the GLGF loop and an Arg residue, through specific hydrogen bonded interactions of its side chain (*e.g.* H372...*T*-2), and through interactions with a pronounced hydrophobic pocket of PDZ3 (*e.g.* F352...*V*0). Residues that contribute to ligand complexation are underlined in the PDZ3 sequence; corresponding residues in the nNOS PDZ domain are also marked. It is probable that the nNOS PDZ domain forms complexes of similar structure with C-terminal peptide ligands. However, the presence of Tyr and Asp residues at positions 77 and 78 in the nNOS PDZ domain (corresponding to His and Glu residues in PDZ3) accounts for its selective recognition of DXV C-terminal consensus sequences rather than (S/T)XV sequence for which PDZ3 exhibits specificity

(*433*)

Fig. 45. Schematic representation of nNOS association with the dystrophin complex at the sarcolemma. Dimeric nNOS binds to the complex by a homotypic interaction between its N-terminal leader PDZ domain and the corresponding domain in α1-syntrophin. The latter is an adapter protein which possesses two modules widely distributed in signalling proteins, namely pleckstrin homology (PH) domains. The first PH domain in the syntrophin is split by an insert that contains the PDZ domain. Dystrophin, which is thought to bind the syntrophin in a region just down stream of its PDZ domain, links transmembrane glycoproteins (dystroglycans, DG) and the actin-based cytoskeleton of the sarcolemma. The glycoproteins of the dystrophin complex mediate the clustering of other proteins of the signalling complex (not shown) such as acetylcholine receptors. The dystrophin complex plays a major role in neuromuscular development and disease [see Bredt *et al.* (*248, 375*) and references therein]

In addition to its possible interaction with G(D/E)XV C-terminal peptide ligands, the nNOS PDZ domain also forms homophilic associations with other PDZ-containing proteins. In this manner the PDZ domain of nNOS has been shown (*375*) to associate with PDZ domains in α1-syntrophin (Fig. 45) and the synaptic proteins PSD-95 and PSD-93 (Fig. 46). The syntrophins are a family of proteins that are predominantly localised at the sarcolemma in skeletal muscle through their binding to the C-terminal domains of dystrophin and homologues; they are particularly abundant at the neuromuscular junction where it is thought that they participate in acetylcholine receptor clustering (*444*). BREDT, PONTING and co-workers (*248, 432*) have discerned significant homology between the nNOS leader sequence and the syntrophin family. It is probable that the α1-syntrophin adapter protein also mediates the association of other proteins with the dystrophin complex in addition to nNOS.

In the brain, association of nNOS is mediated not by the dystrophin complex, but by binding to the synaptic protein PSD-95. A related protein, PSD-93, appears to mediate membrane association of the enzyme in non-neuronal cells of some glandular tissue (*375*). The first two of the three tandem PDZ domains of PSD-95 can bind NMDA receptor subunits (NR2) as well as Shaker-type K^+ channel subunits [see GOMPERTS (*437*) and references therein]. Both types of subunit contain the (T/S)XV C-terminal motif. Curiously, studies have shown (*375*) that the nNOS PDZ domain competes with the terminal (T/S)XV motif of NR2 subunits for binding at PSD-95 PDZ2. However, as the receptor subunit can independently bind to PSD-95 PDZ1, GOMPERTS (*437*) has suggested that the binding of nNOS to PSD-95 is unlikely to preclude NR2 binding. Therefore, the simultaneous interaction of PSD-95, perhaps as an oligomeric complex, with both the NMDA receptor subunits and nNOS is possible. Thus, PSD-95 could serve as a scaffold for the linking of receptors with their signal transduction machinery, leading to appropriate spatial organisation for signalling cascades. The co-localisation of nNOS and NR2 (or other receptors/ion channels) by PSD-95 *in vivo* and the detailed organisation of these possible complexes remains to be determined. A number of other proteins related to PSD-95 [PSD-93, SAP-97, SAP102 and chapsyn (<u>ch</u>annel-<u>a</u>ssociated <u>p</u>rotein of synapses)-110] appear to be equally effective in clustering both NMDA NR2 subunits and Shaker channels [see PONTING *et al.* (*436*) and references therein]. It is unlikely that a single one of these proteins co-localises exclusively with one or other of its ligands. Indeed, clustered channels are known to associate with heteromultimers of PSD-95 and chapsyn-110 (*447*); oligomerisation of these two proteins could itself potentially be mediated by homotypic PDZ-PDZ interactions.

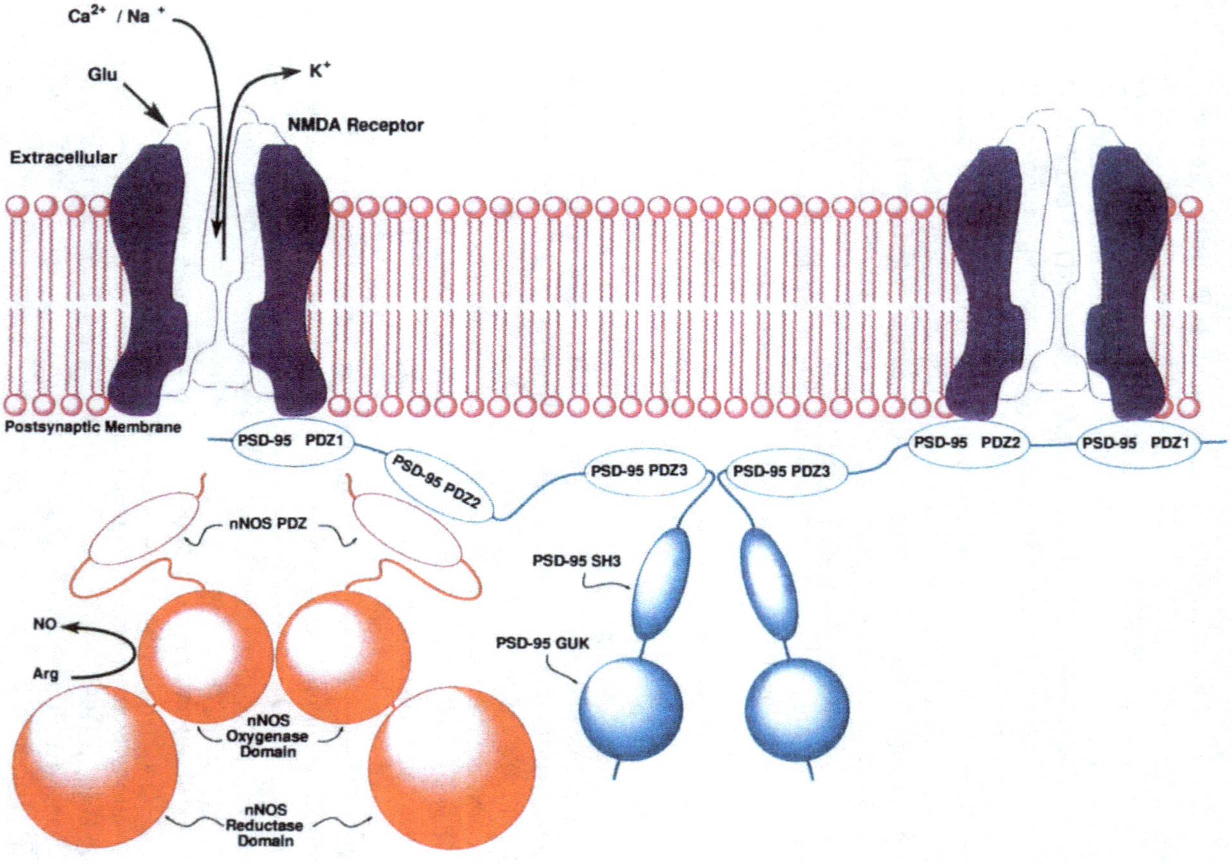
Ca²⁺ / Na⁺
Glu
K⁺
NMDA Receptor
Extracellular
Postsynaptic Membrane
PSD-95 PDZ1
PSD-95 PDZ2
PSD-95 PDZ3
PSD-95 PDZ3
PSD-95 PDZ2
PSD-95 PDZ1
PSD-95 SH3
PSD-95 GUK
nNOS PDZ
NO
Arg
nNOS Oxygenase Domain
nNOS Reductase Domain

The PDZ-PDZ interface in homophilic protein associations probably involves the peptide recognition surfaces of the two interacting PDZ domains as LI *et al.* (*433*) have shown that deletions in nNOS PDZ which abolish binding of C-terminal peptides simultaneously eliminate binding to α1-syntrophin or PSD-93. However, HENDRIKS and co-workers have shown (*443*) that the first 111 residues of nNOS, which incorporate the enzyme's entire PDZ domain, are insufficient for its association with PSD-95 PDZ2. On the other hand, this group found that a larger section of the nNOS leader sequence (residues 1–195) interacted strongly with PDZ2 and, therefore, inferred that the interaction between the two proteins is not of a purely homophilic PDZ–PDZ nature. Although the detailed organisation and role of nNOS in submembranous protein complexes is yet to be revealed, the importance of correct subcellular targeting of the enzyme is underlined by the link between pathophysiological conditions and nNOS-deficient membranes. Thus, aberrant dystrophin genes, which cause nNOS-deficiency in sarcolemmal membranes, are a feature of certain muscular dystrophies (*248, 448–451*). Recently Ross *et al.* (*452*) have speculated on a link between nNOS and Huntingtin disease (HD). HD is characterised by a massive loss of certain neurons caused by an aberrant gene that encodes the huntingtin protein. The Ross group have observed colocalisation of nNOS with a huntingtin-associated protein (HAP-1) in a manner reminiscent of the nNOS/dystrophin complex association; they speculate that altered interactions in the nNOS/HAP-1/mutant huntingtin complex may result in excessive production of NO or its biosynthesis at inappropriate intracellular sites, leading to neuronal damage.

Fig. 46. Schematic representation of the possible association between nNOS and the synaptic protein PSD-95. PSD-95 comprises three tandem PDZ domains, an src homology 3 domain (SH3; a module widely distributed in signalling proteins) and a guanylate kinase homologous domain (GUK) [see PONTING (*436*) and references therein]. PDZ1 and PDZ2 are both able to bind the C-terminal (T/S)XV motif of the modulatory NR2 subunit of NMDA receptors. However, PDZ2 is competitively bound by nNOS through a homophilic PDZ-PDZ interaction between the two proteins. PSD-95 may well be organised into multimeric complexes either with other PSD-95 units or with related proteins of a similar domain organisation such as chapsyn-110, in which case such oligomerisation may also potentially be mediated by PDZ-PDZ interactions. Little is known about the function and protein binding partners of the SH3 and GUK domains in PSD-95. However a family of proteins, the SAPAPs, that binds to the GUK domain has recently been discovered (*445, 446*). The function of SAPAPs is unknown but they may act as adapter units to facilitate the binding of other molecules at the plasma membrane. Co-clustering of nNOS and NMDA receptors by PSD-95 and its homologues *in vivo* and the detailed organisation of these possible complexes remains to be determined

Endothelial NOS is associated primarily (>90%) with membranes in its subcellular distribution (*178, 453*) and in particular with the Golgi complex (*454–456*) and the plasma membrane, where it is localised to the specialised caveolae microdomains (*237, 457–459*). A rationale for this membrane localisation has been suggested in the possible coupling of cell surface receptors or physical stimuli (fluid sheer stress in the blood vessel) to activation of the enzyme (*460*). The association of eNOS with the cell membrane arises in part from myristoylation at the N-terminal sequence of the enzyme, Fig. 47 (*156–158, 379, 424, 461, 462*). Both iNOS and nNOS lack this sequence and site-directed mutagenesis of eNOS G2 has been found to afford a cytosolic eNOS mutant (*379, 424*). The G2A mutant eNOS has been characterised by EPR studies and exhibits an identical heme environment and Arg-binding characteristics to the wild-type enzyme (*309*). The oxygenase domain of NOS is, therefore, probably largely unaffected by its membrane attachment.

Evidence suggests that the membrane association of eNOS is dependent on more than myristoylation alone. For example, Venema and associates (*460*) have shown that eNOS attachment to lipid bilayers involves electrostatic as well as hydrophobic interactions and, in particular, that basic residues within the eNOS CaM-binding domain interact with acidic membrane phospholipids. Thus, deletional mutation of the CaM-binding domain (bovine eNOS residues 493–512) was shown to abolish the enzyme's capacity for membrane attachment and convert eNOS into a cytosolic protein. Moreover, the association with phospholipids was found to inhibit enzyme's catalytic activity by influencing its interaction with CaM. Membrane association, therefore, participates in the regulation of eNOS activity.

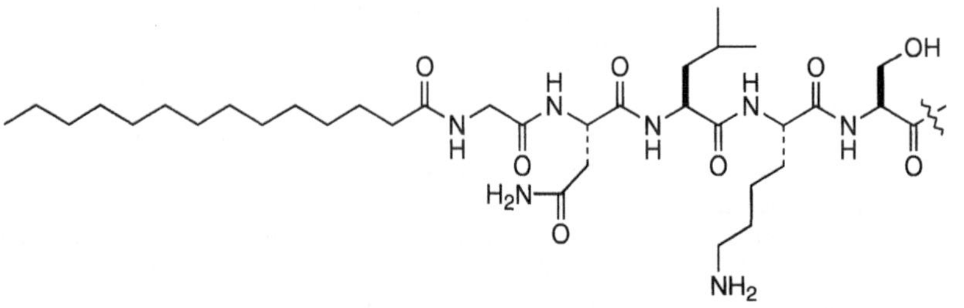

Fig. 47. Endothelial NOS possesses an *N*-myristoylation consensus sequence (MGNLKS) that nNOS and iNOS lack. Myristoylation is a co-translation modification catalysed by *N*- myristoyltransferase. In this process the terminal methionine is removed to allow attachment of myristic acid *via* an amide group to the amino function of G2 [see Robinson and Michel (*463*) and references therein]

In related studies TANIGUCHI *et al.* (*191*) have synthesised peptides encompassing the CaM-binding domain of human nNOS, iNOS and eNOS (Fig. 48). The CaM-binding properties of the eNOS peptide were confirmed and its binding both to CaM and to acidic phospholipids was found to cause a profound structural change in the peptide from random coil to α-helix. The CaM-recognition sequences of iNOS and nNOS were similarly found to bind to phospholipid membranes and, therefore, the NOS CaM-binding domain appears to be able to interact with phospholipids in all three isozymes. The interaction of these peptides with phospholipids is amphiphilic in nature, that is, involving a combination of hydrophobic and ionic interactions, the latter between basic residues in the peptide sequence and acidic phospholipid groups. TANIGUCHI *et al.* also demonstrated that the synthetic eNOS peptide is stoichiometrically phosphorylated by protein kinase C (PKC), leading to a decrease in its interaction with membrane phospholipids. These results are interesting because protein phosphorylation plays a central role in signal transduction pathways regulating many biological processes. Indeed it has been suggested that phosphorylation of eNOS influences translocation of the enzyme from the membrane to the soluble fractions upon suitable stimulation of endothelial cells, a process which may have an important role to play in the function of this enzyme (*188, 189, 192*). CaM-binding domains of other proteins are also known to interact with phospholipids and to be subject to phosphorylation which mediates translocation between membrane and cytosol [see TANIGUCHI *et al.* (*191*) and references therein]. A number of other reports have also described regulation of both of iNOS and eNOS by acidic phospholipids (*464–467*).

In all probability the detailed mechanism which regulates eNOS membrane association, its subcellular localisation and the level of NO synthesis activity of the enzyme is highly complex. For example, the involvement of a polybasic RRKRK motif within the FMN-binding

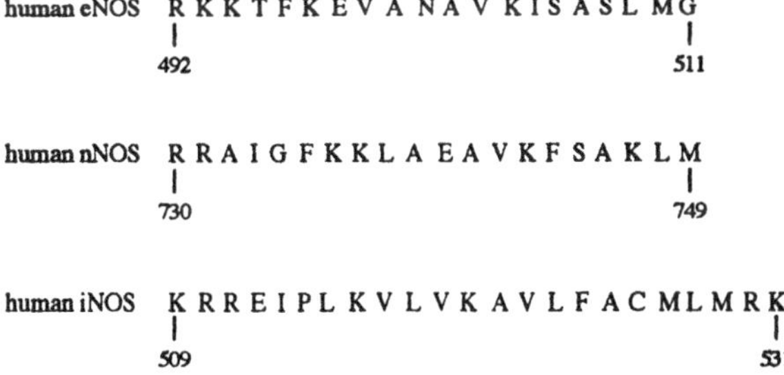

Fig. 48. Primary structure for the CaM-binding domains of human NOS isoforms

sequence has also been considered as a possible determinant of eNOS association with anionic phospholipids, although a deletion mutant lacking this motif retained the capacity for binding to membranes (*462*). Other reports have demonstrated that eNOS is modified by palmitoylation of C15 and C26 at the N-terminus of the enzyme (*193, 380, 463, 468*). Palmitoylation of eNOS appears to be a dynamically regulated process (*193, 418, 469*) and is dependent on prior myristoylation (*380*) which is a co-translational modification and is not dynamically regulated. The importance of palmitoylation to the regulation of eNOS membrane association is somewhat controversial however. Thus MICHEL and co-workers (*193, 463*) have suggested that membrane association is markedly reduced in the absence of palmitoylation and that palmitoylation-depalmitoylation cycles may facilitate translocation of the enzyme between membrane and cytosol. SESSA's group, on the other hand, have reported that palmitoylation of the enzyme is not a critical determinant for its membrane localisation (*380*). Instead these authors argued that other amino acid sequences and hydrophobic factor(s) in eNOS, together with *N*-myristoylation, are primarily responsible for targeting the enzyme to membranes. This group subsequently found (*468*) that myristoylation, possibly in conjunction with interactions of the CaM-binding domain, is sufficient for membrane association and targeting of eNOS to the cell's Golgi complex. Palmitoylation, though having little overall effect on the enzyme's membrane association, was shown to influence its localisation to plasmalemmal caveolae however.* Moreover, NO production in cells transfected with a palmitoylation-deficient eNOS mutant was found to be sub-optimal. SESSA *et al.* pointed out that many signal transducing proteins are targeted to the plasmalemmal caveolae and that perhaps their trafficking to and from these organelles regulates the frequency and magnitude of cellular responses [see SESSA *et al.* (*468*)

---

* In further work SESSA *et al.* (*470*) have recently shown the membrane association determinant of eNOS to be restricted to the first 35 residues of the protein; myristoylation (but not palmitoylation) was confirmed as the primary requirement for membrane association. Palmitoylation is required for correct intracellular targeting of the enzyme and optimal NO production. KELLY and co-workers report (*237*) that eNOS is initially translated in cardiac myocytes as a nonpalmitoylated 150 kDa protein which undergoes post-translational modification to a 135 kDa form. The latter is able to accommodate palmitoylation and is targeted to the sarcolemmal caveolae in these cells. The origin of the additional apparent mass in the 150 kDa form of the enzyme is unknown at present–its predicted mass is 135 kDa based on the mRNA for the enzyme. However, it is clear that post-translational processing of the enzyme, which KELLY *et al.* found to be regulated by protein kinase A-dependent phosphorylation, plays an important role in its subcellular targeting.

and references therein]. Therefore, palmitoylation of eNOS may be responsible for targeting the enzyme near to other signalling proteins to form an efficient signal transduction cascade. In support of this hypothesis the same group have also shown (*190*) that tyrosine phosphorylation of eNOS, which is associated with a reduction in the activity of the enzyme, mediates an interaction with caveolin; the latter is the principle coat protein of caveolae which forms an oligomeric structural scaffold and is integral with the membrane. MICHEL *et al.* have proposed (*471*) that caveolin forms a heteromeric inhibitory complex with eNOS, although it is yet to be established whether these two proteins actually interact directly.* Interestingly, the findings of MICHEL's group suggest that Ca^{2+} levels differentially modulate the association of eNOS with caveolin *versus* CaM. Consequently, the group postulated that eNOS activity is reciprocally regulated by interactions with Ca^{2+} / CaM and caveolin.

In summary, it is likely that the activity and physiological function of both constitutive NOS isoforms is regulated to some degree by membrane association. In the case of nNOS the PDZ domain within the N-terminal leader sequence facilitates clustering with other proteins involved in signalling pathways. Thus, interactions with PSD-95 and syntrophin bring nNOS into close proximity with NMDA receptor channels in neuronal synapses and localise the enzyme to sarcolemmal membranes in skeletal muscle (*248, 375*). Overproduction of NO in the brain mediates injury in cerebral ischemia and animal models of Parkinson's disease (*472–475*). Therefore, despite the importance of NO as a signalling molecule in the nervous system, its biosynthesis must be tightly regulated because of its potential for tissue damage. It is likely that the subcellular targeting of nNOS to membrane regions designed for specific functions reflects the need for careful modulation of the enzyme's activity and perhaps restricts the influence of NO produced to specific targets within a limited microenvironment (*248*).

The endothelial isoform appears to be similarly targeted to specific subcellular regions of the cell membrane and may also be associated with multiprotein complexes that mediate signal transduction. Membrane composition is likely to influence eNOS activity, and findings by DELICONSTANTINOS *et al.* (*476*) support this view. This group have shown that lower cholesterol levels in endothelial cell membranes increase membrane fluidity and up-regulate eNOS activity. Membrane association

---

* See Addendum.

of eNOS arises from myristoylation and palmitoylation of specific residues at the N-terminus of the enzyme, but the CaM-binding domain has also been shown to play a key role in the interaction between enzyme and membrane. The CaM-binding region connects the NOS reductase and oxygenase domains and possesses a basic, amphiphilic nature with an α-helical secondary structure. It is likely that the eNOS CaM-recognition sequence is subject to phosphorylation, which clearly plays an important role in regulating the enzyme.* If so, then this remarkable domain of the enzyme can interact with at least 3 components: CaM, anionic phospholipids and a protein kinase such as PKC. The domain not only regulates electron transfer within the enzyme but also contributes to control of its subcellular compartmentalisation, at least for the endothelial isoform.

10.5. Structure and Topology of the NOS Catalytic Site

10.5.1. Identification of Heme Thiolate Ligand

In 1992 four reports (*136–139, 166*) identified heme as a NOS prosthetic group that is present in stoichiometric quantities and is required for activity of the enzyme. Indeed, NOS catalysis is disrupted by heme iron ligands such as CO (*139*) and simple imidazole derivatives (*477*) which are known to inhibit cytochromes P450 by heme ligation. GIOVANELLI *et al.* (*335*) have shown that CO inhibits not only the overall conversion of Arg to Cit and NO but that both steps in the Arg–NO pathway are disrupted, suggesting that heme mediates both NOS monooxygenations I and II. Early sequence and modelling studies tentatively assigned one particular Cys residue and its surrounding peptide as the likely heme-binding site in the N-terminal domain of the protein (*137, 329*). Shortly afterwards resonance Raman spectra confirmed the axial thiolate ligation of NOS heme (*478*). A low similarity between the heme region of cytochrome P450$_{\text{BM-3}}$ and part of the NOS oxygenase domain has been proposed (*479*) which would fit C672 (rat nNOS) as the axial heme ligand. However, site-directed mutagenesis studies subsequently indicated that C415, C184, C200 and C194 provide the proximal thiolate ligand to heme in rat neuronal (*383–385*), human endothelial (*381*), human inducible (*386*) and murine inducible (*358, 480*) NO synthases respectively. This cysteine is

---

* The number and location of other phosphorylation sites in the enzyme have yet to be ascertained.

References, pp. 144–186

conserved in all NO synthases examined regardless of species. Although some similarity to cytochromes P450 exists in the sequence surrounding C415 in rat nNOS (*329*), the resemblance is modest (*383*). Thus NOS may not be closely related to cytochromes P450 at the structural level, despite possessing typical cytochrome P450 spectral characteristics.

NOS mutants lacking the proximal Cys ligand not only show a loss of heme but also a complete incapacity towards H_4B and substrate binding (*384, 386*), thus indicating that the heme-, substrate- and H_4B-binding sites are closely allied. In other respects, however, these mutant enzymes show little perturbation of protein structure compared to the wild-type enzyme and retain both the tightly bound flavin cofactors and the function of the reductase domain. That the heme- and H_4B-binding sites are in close proximity is further borne out by studies with nNOS and iNOS which show that H_4B directly influences the spectral characteristics of heme (*384, 481*).

10.5.2. Characterisation of the NOS Heme Spin State

In the last few years characterisation of NOS by a number of biophysical methods including UV-visible (*166, 406, 482*), magnetic circular dichroism spectroscopy (*483*), resonance Raman spectroscopy (*478, 481, 484*) and electron paramagnetic resonance (EPR) (*166, 309, 395, 398, 485, 486*) has provided insights into the structure-function relationship and redox behaviour of the enzymes. The latter technique in particular has been used to provide information about heme symmetry, identify heme ligands, probe structural perturbation caused by different axial ligands and contribute to an understanding of the NOS heme iron's electronic structure. Thus, the heme groups of all three mammalian NOS isoforms have been characterised by their CO difference spectra (*137, 139, 166, 381, 480, 482, 487, 488*) and by EPR studies (*166, 309, 398*). Most cytochrome P450s contain mainly six-coordinate, low-spin ferric heme in the resting state whereas a number of reports (*166, 478, 482, 489*) attribute a predominantly high-spin state to NOS, with optical and EPR characteristics that are similar to high-spin heme in cytochrome P450 complexes. However, the heme spin state of NOS appears to be dependent on the source of the enzyme. Indeed the proportion of low-spin heme varies from preparation to preparation (*166, 309, 398, 485*), with several publications recording a predominantly or exclusively low-spin heme resting state for the enzyme, particularly when the latter is expressed in H_4B-free bacterial systems (*385, 422, 487, 489–492*). The presence of H_4B and

Arg shifts the equilibrium in favour of the high-spin state; thus bound H_4B and substrate appear to limit solvent accessibility to the distal heme pocket. TSAI and co-workers (*309*) have observed two distinct heme species in the low-spin component of resting eNOS and attributed this to the presence of two different distal ligands, possibly H_2O and OH^-. The influence of substrate and inhibitor binding to NOS has also been investigated by EPR studies (*309, 395, 398, 485, 493*).* Substrate binding has been shown (*398*) to take place outside the first coordination shell of the heme iron, and both Arg and NHA produce exclusively 5-coordinate high-spin heme complexes with NOS (*309, 485*). The spectral perturbation brought about by substrate binding to NOS has been taken as direct evidence implicating heme as the reaction centre for initial substrate binding and subsequent oxidation (*482*). Cytochromes P450 such as $P450_{cam}$ possess a notably hydrophobic distal heme pocket (*266*). However, analysis of the distal heme environment in eNOS, made by comparison of the heme characteristics of eNOS-ligand complexes with those of other hemoproteins, suggests that the catalytic site of NOS more closely resembles chloroperoxidase than a cytochrome P450 (*309*). The relatively hydrophilic character of the distal heme environment in NOS is not unexpected considering that it interacts with polar substrates, Arg and NHA.

10.5.3. Structure of the NOS Catalytic Site

Several studies with NOS inhibitors have sought to probe the spatial relationship of Arg and H_4B binding sites with respect to the heme centre; the inhibitory properties of imidazole derivatives, which bind by direct coordination to the heme iron, have been particularly useful in this regard (*477, 482, 488, 494–498*). Initial investigations with rat and porcine brain NOS afforded conflicting results regarding the relationship between imidazole and substrate binding in the catalytic site, some suggesting a competitive binding mode and others noncompetitive binding (*496–498*). More recently, the availability of larger quantities of recombinant human inducible NOS (rH-iNOS) has allowed WONG and

---

* MASTERS *et al.* (*395, 486*) have shown that Arg, NHA, Cit and Arg analogue inhibitors of NOS produce distinctive distortions in the ferriheme iron ligand geometry which can be detected in the EPR spectra of NOS-ligand complexes. This allows spectroscopic fingerprinting for occupation of the substrate binding site. The sensitivity of EPR spectra to different Arg derivatives and analogues provides a powerful tool for analysis of interactions between the O_2-binding heme centre and the Arg-recognition site of the enzyme.

References, pp. 144–186

the team at Merck Research Laboratories (*488*) to undertake a thorough kinetic and spectrophotometric study of NOS inhibition by imidazole and substituted imidazole derivatives. This work has shown that imidazole inhibitors bind reversibly and compete with the substrate for a mutually exclusive site.

Structure-activity relationship studies within the imidazole inhibitor series demonstrated a potent drop in inhibitory capacity when 2- or 4-substituents were present, presumably due to unfavourable steric interactions with the heme moiety. On the other hand, a variety of N^1-substituents were tolerated, including alkyl, benzyl and aryl groups. Interestingly, 2-(1-imidazolyl)ethanesulfonic acid was 200-fold less potent as an inhibitor than the corresponding ethanamine (Fig. 49). Such a difference could infer a favorable interaction between the basic amine of the latter and an acidic residue in the protein, a residue that would be of significance in view of its potential contribution to a binding motif for the Arg guanidino group. Indeed a conserved glutamic acid residue in the catalytic site of NOSs has since been shown to interact with the substrate guanidine function (*298, 299*). However, WONG and co-workers observed that imidazoles carrying lipophilic N^1-substituents (e.g. butyl) exhibit similar potency to 2-(1-imidazolyl)ethanamine, suggesting that a specific ionic interaction between the protonated amino group of this inhibitor and a carboxylate residue in the catalytic site is unlikely.

The Merck group also discovered that imidazole and *N*-phenylimidazole competitively inhibit H_4B binding in addition to Arg binding (*488*). This confirms that the H_4B binding site as well as the Arg binding site are in very close proximity to the heme centre and once again raises the question of the catalytic role of the pterin cofactor. Variations in the inhibition parameters of imidazole with NOS enzymes from various sources were taken to indicate that slight differences exist in the spatial arrangement of these binding sites with respect to heme. Further studies using photoaffinity probes such as 1-(4-azidophenyl)imidazole are in

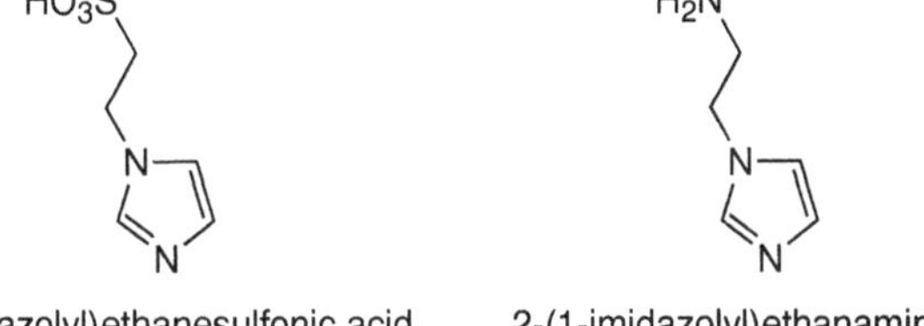

Fig. 49. Imidazole inhibitors used by WONG *et al.* (*488*) to probe the catalytic site of NOS

progress at the Merck Research Laboratories and could usefully identify residues that contribute to the catalytic site.*

Ortiz de Montellano and associates have probed the NOS catalytic site using aryldiazene methodology (*499, 500*). Aryldiazenes (Ar–N= NH) react with the enzymes to produce a σ-bonded aryl-iron complex with a distinctive absorption in the UV-visible spectrum; ferricyanide oxidation triggers migration of the aryl group to the porphyrin nitrogen atoms. Two elements of these studies provide useful information about the topology of the NOS catalytic site: firstly, analysis of the rate and extent of aryl-iron complex formation in the presence and absence of substrate and cofactors (Ca^{2+}/CaM, H_4B); and secondly, the ratio of regioisomeric *N*-aryl migration products formed, since steric obstructions in one region of the catalytic site and not in another cause preferential aryl group migration in the least hindered direction. In each of the three mammalian NOS isoforms aryl group migration proceeded predominantly to pyrrole ring D (refer to Fig. 9), indicating that the region above this ring is most open, although none of the rings is completely masked. Indeed, the active site topologies of the three isoforms are broadly similar, although the use of sterically more demanding diazene probes uncovered some difference between the isoforms. The ceiling height directly above the heme iron is insufficient to permit formation of a *p*-biphenyl-iron complex in all 3 isoforms, indicating a roof of lower than 9.9 Å. On the other hand, the active sites of nNOS and iNOS, but not eNOS, are able to accommodate a 2-naphthyl iron ligand which requires a head space of $\sim$7.1 Å.

The extent of iron complexation on treatment with phenyldiazene is significantly decreased in the presence of Arg. This is not surprising as the Arg guanidine function is known to bind close to the plane of the heme, and must do so in order to undergo oxidation. In the presence of Ca^{2+}/CaM the two constitutive isoforms, which are sensitive to levels of this cofactor, exhibited slight changes in the rate and extent of the phenyldiazene reaction as well as in the population of *N*-phenylporphyrins formed on ligand migration. These observations suggest that Ca^{2+}/ CaM binding to NOS induces some small conformational change in the catalytic site, but it is unclear what the relation of this change is to the enhancement of electron flux triggered by the cofactor. The influence of H_4B on the phenyldiazene reaction of the three isoforms was generally

---

* Inactivation of the human enzyme by this photoaffinity probe has been accomplished, but non-specific labelling has predominated over specific labelling in early studies. (Personal communication from Dr. K. Wong, Merck Research Laboratories, New Jersey).

consistent with a restriction in the overall cavity size of the active site. Of more importance was a significant decrease in the phenyl shift to pyrrole D with compensatory increases at the other rings, and particularly pyrrole B (the ring located diametrically opposite to ring D). These observations may indicate that H_4B binds above pyrrole ring D as spectroscopic studies have shown that the heme- and H_4B-binding sites are closely linked (*488, 501*). However, it also possible that H_4B binding at some other point near the catalytic site produces a conformational change that brings protein residues into closer proximity with the porphyrin D ring.

Tsai, Chen and co-workers (*502*) have also examined the spatial relationship between the Arg and heme binding sites of eNOS through a study of spectral perturbation with Arg analogues and heme ligands; these investigations have yielded useful information regarding the physical dimensions of the distal heme ligand site. Two groups of ligands were identified: one group, type II ligands, that binds by direct ligation of the heme iron and results in a 6-coordinate low-spin complex; and a second group, type I ligands*, that binds in the heme distal pocket without direct ligation of the metal centre. These groups are exemplified by imidazole and Arg respectively. Type I ligands displace the original axial heme ligand present in the low-spin population of resting NOS, to produce a 5-coordinate high-spin complex.

The binding of both imidazole and Arg was found to be rather a slow process and similar kinetic characteristics suggested that both ligands access the binding site, which is deeply buried in the protein, through a common passage. It is assumed that steric interactions prevent Arg and other relatively large ligands containing a guanidine motif from directly ligating the metal centre. In the imidazole series a weaker binding affinity of 2-methylimidazole was attributed to steric strain imposed by the presence of a methyl substituent. The bulkier pyridine and pyrimidine ligands were found to exhibit bimodal behaviour, that is, the parent unsubstituted molecules are type II ligands that coordinate to heme iron, but their bulkier derivatives, 4-methylpyrimidine and 3,5-dimethylpyridine, are type I ligands. These delimiting ligands were used to gauge the dimensions of the distal heme pocket (Fig. 50). In considering the estimated dimensions of the catalytic site protein dynamics should also be taken into account because the site dimensions

---

* Types I and II strictly refer to the characteristic spectral changes induced by the ligands on complexation with the enzyme; these changes fall into 2 classes and reflect the mode of ligand binding (*482*).

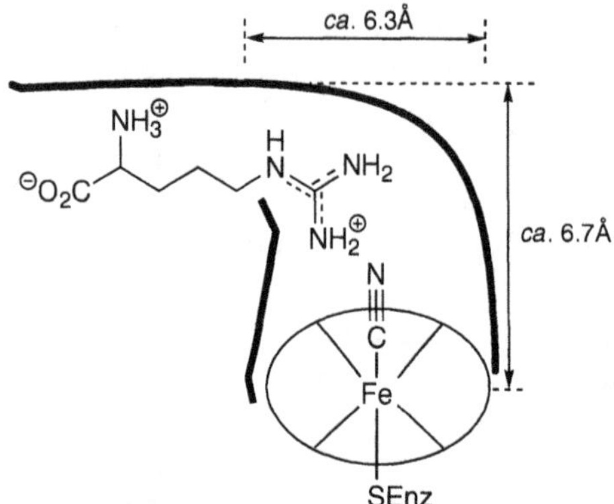

Fig. 50. Schematic representation of the NOS catalytic site. Arg binds to NOS through a defined channel. The dimensions of the immediate distal heme pocket have been estimated by Tsai et al. (*502*). A small type II ligand, cyanide, can be accommodated in the catalytic site without disruption by Arg binding. Larger type II ligands such as imidazole compete with Arg for overlapping regions of the distal heme pocket. Other work by Ortiz de Montellano et al. (*500*) suggests that the head space above the porphyrin iron lies in the region of 7.1–9.9Å

will most likely be influenced by changes in protein conformation to accommodate the different ligands.

The binding volume of type II ligands overlaps with the binding volume for guanidine containing type I ligands; thus Arg and imidazole bind to NOS competitively. The *only* type II ligand studied that was small enough to coordinate to NOS heme without conflict with Arg was cyanide, which underlines the close proximity of the substrate guanidine functionality to the heme plane in the enzyme-substrate complex. Tsai et al. concluded that the guanidino nitrogen is suitably orientated towards the distal heme coordination site to allow reaction with the perferryl oxoiron intermediate by hydrogen abstraction and oxygen rebound.

In an extension of the above studies Tsai and co-workers (*309*) have defined a set of direct heme-coordinating type II ligands for which a cognate group of type I non-coordinating ligands with similar functionality exists (Fig. 51). Imidazole was found to bind in two distinct orientations relative to the heme plane so as to produce a heterogeneous EPR spectrum. A pH reduction from 8.0 to 6.9 resulted in the attenuation of the signal from one of the eNOS-imidazole complex forms, possibly due to the elimination of a hydrogen bond stabilising that particular bound orientation.

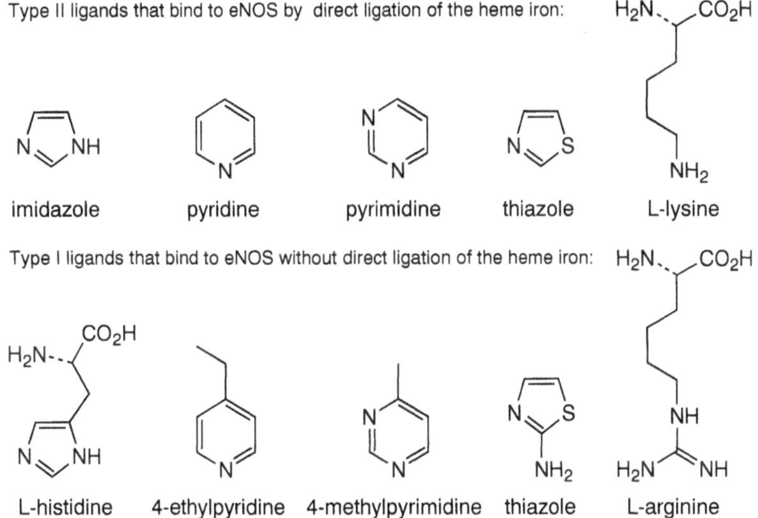

Fig. 51. Cognate ligand groups with different modes of NOS binding have been identified by TSAI *et al.* (*309*)

10.5.4. Identification of Residues that contribute to the Catalytic Site*

Various reports have attempted to throw light on the specific interactions between NOS and Arg that contribute to substrate binding (*503, 504*). Studies with Arg analogues and heme ligands have demonstrated that the substrate guanidino function is critical for molecular recognition and binding to NOS (*502, 505, 506*), but a variable degree of importance has been credited to the positive charge it carries in the protonated guanidinium state (*308, 507–509*). More recently CHEN, TSAI and co-workers (*298*) have identified a glutamate side chain in the three human isoforms that forms a specific ionic interaction with the substrate guanidinium moiety. This discovery followed localisation of the nNOS Arg-binding site to residues 558–721 in the N-terminal domain of the protein (*510*). CHEN *et al.* made sequence alignments for the three isoforms and identified a 40-residue segment which displayed high similarity across all 3 isoforms, Fig. 52. Site-directed mutagenesis studies were then used to probe the role of several acidic residues within the 340–380 segment of human eNOS (*298*). E342I and E377I mutations had no significant effect on enzyme activity, whereas E347L, E361L/Q and D369I mutations abolished the Cit-forming capacity of eNOS. Of

---

* See Addendum.

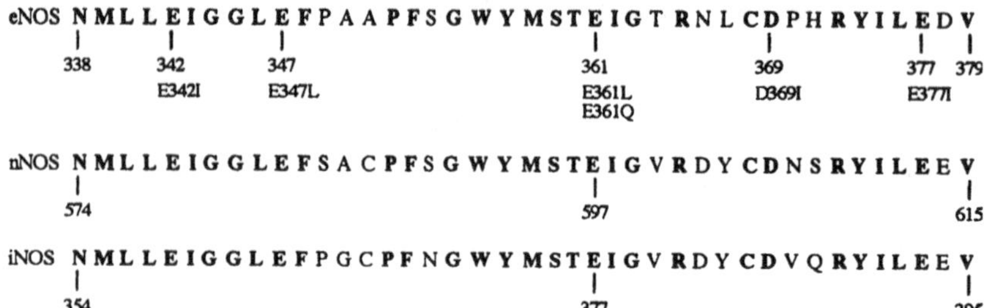

Fig. 52. Human NOS primary structure alignment for endothelial (*155*), inducible (*163*) and neuronal isoforms (*161*) in the putative Arg binding region. Mutations used by Chen *et al.* (*298*) to probe the role of acidic residues are written under the endothelial sequence

this latter group only the E361L/Q mutants preserved the characteristics of wild-type eNOS optical spectra, and these mutant enzymes were found incapable of binding labelled Arg. The E361L/Q mutants were further found unable to bind 2-aminothiazole or acetylguanidine, compounds that are known to bind to the catalytic site of wild-type NOS and mimic the substrate guanidino group but lack its α-amino acid functionality. These results, then, demonstrate that the γ-carboxylate of E361 interacts directly with the substrate's guanidinium functionality rather than its α-ammonium group and that the anionic character of this residue is essential to substrate binding. Also consistent with the polar nature of the NOS heme distal pocket was the observation that the binding affinity of wild-type eNOS for Arg in the presence of NaCl showed a progressive fall off with increasing ionic strength of the medium. The salt bridge between E361 and the Arg guanidinium moiety is thought to provide the major stabilisation energy of the enzyme-substrate complex, Fig. 53.

Chen *et al.* further found resting wild-type eNOS, which contains both high and low-spin heme, to be converted entirely into a high-spin state upon addition of Arg. Arg-binding also increased by 2-fold the NADPH oxidase activity of the enzyme, probably by increasing the heme redox potential in the high-spin eNOS-Arg complex and so facilitating electron transfer to heme (*406*). These observations accord with the catalytic mechanism for NOS monooxygenation I presented in Fig. 19. Not surprisingly the E361Q/L mutants showed no enhancement of NADPH oxidase activity in the presence of Arg, although in other respects they exhibited similar optical and EPR spectra to resting wild-type eNOS. The rate of electron transfer from the NOS reductase domain to cytochrome *c* was unaffected by E361 mutation and, as for the wild-type enzyme, enhanced 10-fold by Ca^{2+}/CaM. A slightly higher proportion of low-spin heme was observed in the E361L mutant, and this

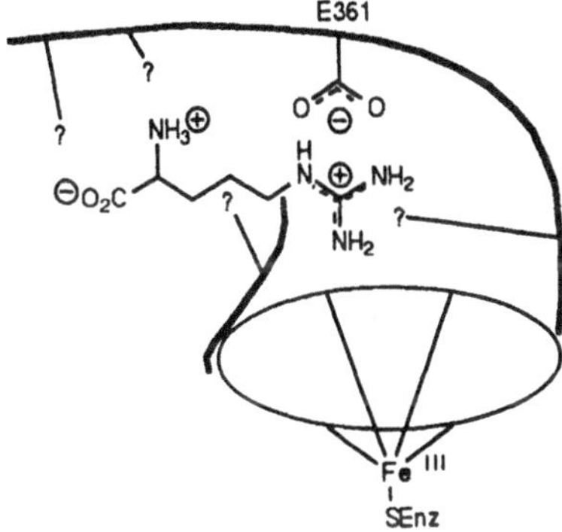

Fig. 53. The major stabilisation energy of the eNOS-substrate complex is provided by a charge interaction between E361 (human eNOS) and the Arg guanidinium moiety (*298, 299*). The nature of additional enzyme-substrate binding interactions, for example to the α-amino acid functionality, is unknown at present

difference was attributed to slight changes in the distal heme allowing greater solvent accessibility and so a higher proportion of enzyme carrying H_2O/OH^- at the sixth heme position. It is interesting that synthetic porphyrins carrying carboxylic functionality close to the *meso* positions are not competent as monooxygenation catalysts because O_2 activation induces rapid attack by the carboxylic group on the *meso* centres (*511, 512*). Such observations may help to define the structure of the NOS catalytic site by suggesting probable exclusion zones which the carboxylate functionality of E361 is unlikely to occupy.

Related studies by STUEHR'S group (*299*) have reinforced the results obtained by CHEN *et al.* Alanine screening mutagenesis of several conserved acidic residues within the isolated murine iNOS oxygenase domain identified two mutations, E371A* and D376A, that selectively caused defects in the Arg binding capacity of the enzyme. However, whereas the former mutation resulted in total loss of substrate binding capacity, the latter substantially attenuated but did not completely abolish substrate binding. Moreover, D376 is not completely conserved across all NOS isoforms. STUEHR *et al.* proposed that the carboxylate of E371 (murine iNOS) interacts with the substrate N^δ and one of the terminal N^ω nitrogens of Arg. This type of interaction features in the binding of Arg by arginase (*513*) and in nature such protein-Arg interactions are far more prevalent than those involving bridging of a carboxylate to both the N^ω nitrogens of Arg. This hypothesis is further supported by binding studies with Arg analogues (*308*) wherein loss of

---

* Murine iNOS E371 is homologous to E361 in human eNOS.

one N^{ω} nitrogen is tolerated but substitution of the N^{δ} nitrogen abolishes high binding affinity. A conservative E371D mutation maintained the Arg binding ability of the enzyme and its catalytic activity.

Significantly, agmatine, the decarboxylated derivative of Arg, and N^{G}-hydroxyagmatine both support NO generation by NOS (*514*), suggesting that the Arg α-carboxylate is not absolutely essential to substrate binding. Moreover, studies of NOS subunit association, which for iNOS is promoted by the presence of substrate, suggest that the α-amino acid component of Arg is non-essential to dimer assembly (*493*). Indeed, simple guanidine analogues such as *N*-hydroxyguanidine effectively promoted dimer assembly by binding to the substrate-recognition site. Thus, the dominant binding interactions between enzyme and substrate most likely involve its guanidine moiety rather than the α-amino acid portion of the molecule.

10.5.5. Complementation Analysis with NOS Heterodimers

Studies with iNOS heterodimers, composed of one native NOS subunit complexed with a modified subunit, have also thrown light on the structure of the NOS catalytic site (*355, 358, 515, 516*). For example, researches in Nathan's laboratory (*358*) showed that point mutation of the proximal Cys thiolate residue in murine iNOS (C194A) abrogates heme binding and results in a monomeric heme-free enzyme. Although the C194A mutant protein is unable to form a homodimer, hetero-dimerisation does occur with the native full-length iNOS subunit. However, this heterodimer, possessing a single heme prosthetic group, is catalytically inactive with respect to NO synthesis. In contrast, a heterodimer between the native iNOS subunit and a truncated subunit lacking the reductase domain retains the capacity for NO synthesis. Thus, a single functional reductase domain within the dimeric NOS structure is sufficient to support NOS catalysis, whereas two heme prosthetic groups are obligatory for activity. On these grounds it was proposed (*358*) that NOS may contain a single active site in which both hemes participate. Alternatively the enzyme may possess two independently functioning catalytic sites organised such that the structure of each is reciprocally maintained by the presence of bound heme in the other subunit.

A related complementation analysis study by Stuehr *et al.* has interpreted results in favour of the dual-site NOS model (*516*). In this work a heterodimer composed of a normal iNOS subunit and a truncation mutant lacking the reductase domain was characterised. NO synthesis occurred at half the rate of that catalysed by the full-length homodimer and, significantly, only one of the two heme groups was subject to reduction. In homodimeric NOS, then, it is possible that each reductase domain acts as a dedicated source of electrons for one of the two heme groups. This model is consistent with an enzyme possessing two independently functioning catalytic sites rather than a single site to which both hemes contribute.

More evidence in favour of the dual-site NOS model has been provided by GORREN, MAYER and co-workers (*517*). This group has found that dimeric nNOS possesses two different affinity states for its H_4B-binding sites (*501, 517–520*). Thus, the dimeric enzyme, though capable of binding two equivalents of H_4B, only tightly binds one as a prosthetic group (*518*). Indeed the affinity for H_4B at a single site in dimeric nNOS is very high, whereas a second equivalent of the cofactor binds only comparatively weakly (*501*). This suggests that NOS forms a symmetrical dimer with two identical H_4B- and Arg-recognition sites, but exhibiting high anticooperativity between the two H_4B-binding sites (*520*). In contrast the binding of H_4B and Arg exhibits positive cooperativity; thus, Arg binding reduces the rate of H_4B dissociation (*519*). GORREN *et al.* (*517*) have recently probed the relationship between the heme / H_4B / Arg-binding sites within the dimer subunits using thiol heme ligands. Dithiothreitol (DTT) was found to bind with high affinity to dimeric nNOS containing one equivalent of bound H_4B but, whereas DTT bound heme in the H_4B-free subunit with high affinity, binding in the H_4B-containing catalytic site was weak. This probably reflects the close proximity of H_4B and heme in the catalytic centre so that the presence of H_4B excludes the DTT ligand. Preincubation of the enzymes with Arg abolished DTT binding in the H_4B-containing site altogether while having little effect on thiol binding to heme in the H_4B-free subunit. Thus Arg binding must occur at the H_4B-containing site and models which assume stimulation of Arg binding in one subunit by the binding of H_4B in the other can be excluded. Moreover, models that involve a single active site formed jointly by both oxygenase domains can be discounted. Finally, the group concluded, from comparative analysis of DTT binding to the enzyme and inhibition of Cit formation, that the H_4B-free subunit does not participate in NO production.*

----

* This issue is still debated, however, because several studies that have evaluated binding stoichiometries of cofactors have found values significantly less than 1 mol/mol of subunit. For example, stoichiometries have been determined in the following ranges: 0.3–0.6 mol NNA/mol of subunit (*374, 489, 521*), 0.06–0.5 mol H_4B/mol subunit (*144, 146, 374, 522*), and 0.4–0.6 mol heme/mol subunit (*139, 374, 383, 523*). This has led to recent speculation that the catalytically active form of NOS may be an asymmetric dimer that contains a single catalytic site, possibly spanning the dimer interface (*374*).

10.5.6. Summary

In order to make further progress in elucidating the mechanism of NOS catalysis it will be essential to throw light on the interaction between the enzyme and its substrate. An understanding of the catalytic site will also assist the development of therapeutically important NOS inhibitors. In particular it will be useful to uncover any differences that exist between the substrate binding sites of the three NOS isozymes which might be exploited for the development of isoform selective NOS inhibitors. A comparison of NOS to other Arg-binding proteins has shown no significant sequence homology (*159*). Moreover, the lack of a 3D structure and absence of significant sequence homology between the NOS oxygenase domain and known cytochromes P450 has made it difficult to identify residues and construct a model of the distal heme pocket responsible for substrate binding. However, a number of groups are currently working towards crystallisation of the separate NOS reductase and oxygenase domains of the three isoforms for X-ray diffraction studies; the first X-ray structure is likely to be forthcoming within a matter of months.* The results of these studies are expected to resolve many of the uncertainties surrounding the structure of the NOS catalytic site. Preliminary X-ray diffraction analysis of CPR from rat liver has already been reported by Masters *et al.* (*524*) and the future emergence of a detailed structure for this protein should throw light on the structure and function of the NOS reductase domain. In addition to these studies, we (DRA unpublished results) and others (*525*) are currently investigating the synthesis and function of model porphyrin systems that mimic NOS. Studies with these models should lead to a clearer understanding of the mechanism of Arg oxidation by NOS and could potentially lead to the development of electrocatalytic probes for the clean, highly-regulated and -directed generation of NO in physiological systems.

10.6. Role of the Biopterin Cofactor in NOS Catalysis

10.6.1. NOS and Pteridine-Dependent Hydroxylases: Dissimilarity in the Catalytic Role of H_4B

In 1989 two groups identified a link between H_4B and the activity of macrophage iNOS (*147, 148*); a similar connection with nNOS (*149*) and

---

* See Addendum.

eNOS (*178*) was demonstrated soon afterwards. Initial studies suggested that H_4B may positively modulate NO biosynthesis without actually being an essential cofactor because nNOS purified to homogeneity exhibited an apparently pteridine-independent basal activity that increased about 4-fold on addition of H_4B (*149*). Around the same period a number of other reports described the isolation of apparently H_4B-independent NOSs (*153, 526, 527*). It subsequently became evident, however, that considerable amounts of H_4B remain tightly bound to purified NOS (*144, 151–153, 172, 522, 528, 529*), which accounts for the enzyme's basal activity in the absence of added H_4B. The importance of H_4B to NOS activity is further underlined by several disclosures that demonstrate a coupling between cytokine-induced pteridine biosynthesis and iNOS induction (*150, 154, 254, 530–535*).

The absolute dependence of NO biosynthesis on H_4B is, therefore, well established although the precise role of this cofactor in NOS catalysis remains something of a mystery at present and requires clarification. In principle, either of the monooxygenations catalysed by NOS could be mediated by H_4B rather than by heme. Hydroxylations catalysed by the aromatic amino acid hydroxylases are, for example, mediated by this cofactor [for reviews see (*174, 175*)]. The aromatic amino acid hydroxylase family, of which there are three members (the phenylalanine, tyrosine and tryptophan hydroxylases), mediate catabolism of phenylalanine and biosynthesis of catecholamines and serotonin. The catalytic cycle of these enzymes. Fig. 54, is thought to involve formation of a 4a-hydroperoxy-H_4B species which participates in substrate hydroxylation at a reduced, non-heme iron catalytic centre, possibly by way of an oxoiron or peroxoiron intermediate, to afford the hydroxylated aromatic amino acid product and 4a-hydroxy-H_4B. As shown, the carbinolamine by-product is recycled *via* quinonoid dihydrobiopterin (qH_2B) to H_4B by the action of two enzymes. Prior to the discovery of a heme cofactor, a similar role for H_4B in the NOS catalytic cycle seemed plausible (*144, 147, 323*). Indeed, 4a-hydroperoxy-H_4B can be envisaged as substituting for the putative peroxoheme intermediate in NOS monooxygenation II. However, several considerations suggest that the role of H_4B in NOS catalysis is dissimilar to its role in aromatic amino acid hydroxylation: firstly, there is the functional analogy between NOS and the cytochrome P450 system and the fact that NOS is inhibited by classical heme ligands such as cyanide, CO and imidazole; secondly, cytochrome P450 enzymes can themselves mediate hydroxylation of guanidine-containing substrates (*331, 332*) as well as oxidation of guanidoximes (*328, 329*) to ureas and NO; and thirdly, NO biosynthesis is dependent on NADPH consumption whereas aromatic amino acid hydroxylation occurs in the absence of NADPH (although this factor does participate in recycling of the pterin by other enzymes).

10.6.2. An Allosteric Role for H_4B in NOS Catalysis

Initially Giovanelli et al. (536) considered catalytic recycling of H_4B in the manner characteristic of the aromatic amino acid hydroxylation cycle unlikely as a proportion of H_4B does not readily dissociate from NOS. They concluded that the H_4B cofactor of NOS was more likely to have an allosteric than a catalytic role. Indeed, radioligand studies with the Arg analogue NNA and H_4B found that binding of the Arg analogue reduced the dissociation constant of H_4B and, conversely, the presence of H_4B converted the Arg-binding site into a high affinity state (501). This mutual enhancement of binding affinities (positive cooperativity) between substrate and pteridine binding sites points towards an allosteric interaction between the two sites. Moreover, the assembly of murine

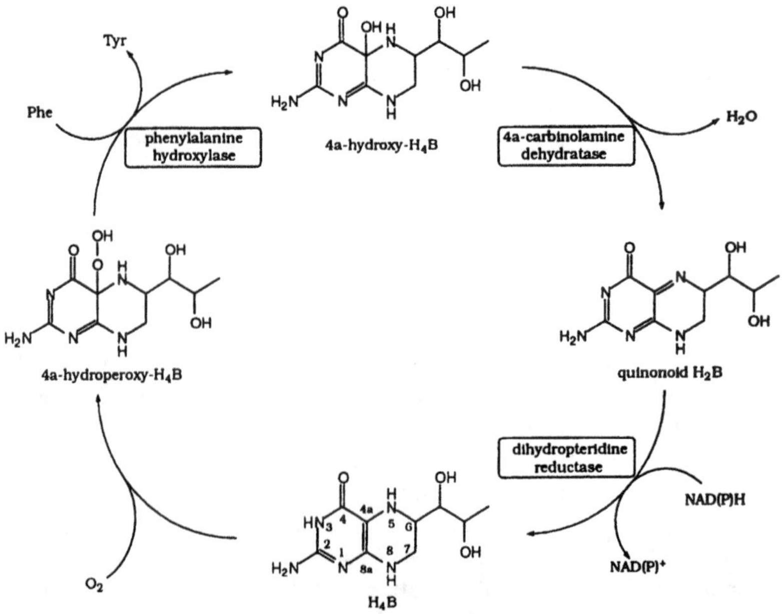

Fig. 54. Catalytic cycle for phenylalanine hydroxylation. Substrate hydroxylation is mediated at a reduced, non-heme iron centre in phenylalanine hydroxylase (PAH) and proceeds with consumption of the 4a-hydroperoxy-H_4B intermediate and formation of 4a-hydroxy-H_4B as a by-product. The latter is subjected to dehydration by 4a-carbinolamine dehydratase, affording quinonoid H_2B which is recycled to H_4B by dihydropteridine reductase

macrophage iNOS subunits into the catalytically active dimeric enzyme has been shown to require the presence of H_4B in addition to heme and Arg (*172*). This finding is also consistent with an allosteric role for H_4B, but such H_4B-dependency is apparently not uniform across the three NOS isoforms (*vide infra*, Section 10.7).

10.6.3. A Redox Role for H_4B in NOS Catalysis

The involvement of H_4B as a stoichiometric reactant or a redox-active cofactor has been difficult to prove because oxidation and recycling of tightly bound H_4B has not been detected directly. One report (*537*) has provided tentative, indirect evidence suggesting that H_4B recycling may be required for NO synthesis, and internal recycling of NOS-bound H_4B is another possibility that cannot be excluded at present. Another key problem which has hampered progress towards an understanding of this cofactor's role in NOS catalysis is the variable H_4B stoichiometry associated with NOS isolates from different sources. However, studies with the neuronal isoform (*518*) have recently begun to shed light on this latter issue and suggest that dimeric NOS contains one tightly bound H_4B together with a site available for binding a second equivalent of the cofactor more losely (*vide supra*, Section 10.5). Moreover, there is now a growing body of evidence which supports a redox-active role for H_4B. Thus, the oxidised pteridine, 7,8-dihydro-biopterin (7,8-H_2B), has been found to bind with fairly high affinity to NOS, but without supporting catalytic activity (*501*). Were the role of H_4B purely one of an allosteric effector then the dihydrobiopterin ought to stimulate NOS activity. A variety of pteridine analogues have been used to probe H_4B function and, although H_4B is the most active cofactor, some support a significant level of NOS activity. For example, (6*S*)-H_4B increases nNOS activity to the same extent as the natural (6*R*)-stereoisomer (about 4-fold), although it must be added at 60-fold higher concentrations to achieve that effect (*501*). In general, however, only redox-active H_4B analogues stimulate NOS activity. Thus, the 6-methyl-5-deaza analogue, for example, inhibits rather than activates the enzyme (*152*).

More recently GIOVANELLI *et al.* (*538*) have demonstrated that NOS itself catalyses an NADPH-dependent reduction of qH_2B to H_4B. This is potentially significant because of the cycling of H_4B and qH_2B in the aromatic hydroxylation pathway. However, reduction of qH_2B was shown to occur on the NOS reductase domain, remote from the high-affinity H_4B-binding site in the oxygenase domain. Therefore, pteridine recycling by NOS, if indeed it does occur, could involve trafficking of the cofactor between the reductase domain binding site and the high affinity H_4B-binding site either by dissociation from the enzyme or by a direct interdomain route. However, reduction of qH_2B took place at concentrations which were significantly higher than the concentrations of H_4B required for full stimulation of NOS. The possibility that H_4B is cycled while fixed at the oxygenase binding site has also not been excluded. The GIOVANELLI group has provided further evidence which supports a redox-active role for H_4B by demonstrating that NOS can catalyse substoichiometric oxidation of Arg to NHA in the absence of NADPH (*335*). This reaction involves O_2 consumption and suggests that H_4B may possibly constitute a reductant for O_2 activation. SCHMIDT *et al.* have also recently invoked a redox function for H_4B by showing that a series of pteridine analogues blocks the NADPH diaphorase activity of NOS (*539*). However, H_4B may not normally be involved in the reductive activation of O_2 by NOS since H_2O_2 generation by the neuronal isozyme in the absence of Arg was pteridine-independent, and reconstitution of pteridine-deficient nNOS with H_4B has no effect on reaction rates (*345, 540*).

If, as seems to be the case, H_4B is not normally directly involved in reductive activation of O_2 then what is the redox role of this cofactor? GIOVANELLI *et al.* (*536*) have speculated that H_4B may be required, perhaps only at substoichiometric levels, to keep some unspecified group(s) in a reduced state, and possibly to reactivate NOS after occasional inactivation during turnover. Interestingly, studies by IGNARRO *et al.* (*541*) suggest that H_4B may play a protective role by limiting NO-mediated inhibition of NOS, which occurs at least in part due to ligation of the heme iron. These investigators found H_4B to be uniquely highly effective in maintaining NOS activity and speculated (*542*) that H_4B may be capable of regenerating the active enzyme.

The role of H_4B in NO biosynthesis, however, is potentially complicated by its implication in the transformation of NO into peroxynitrite (*543*). This appears to proceed by autoxidation of H_4B in the presence of O_2 to produce superoxide, which in turn reacts rapidly with NO to form peroxynitrite. The physiological relevance of this interaction between H_4B and NO is not entirely clear and obviously depends on the levels of H_4B and NO being produced and the presence of other factors such as SOD which remove superoxide. MAYER'S group (*543*) for example, has shown that high levels of SOD are required to detect NO during NOS turnover. While this is consistent with the removal of superoxide generated by autoxidation of H_4B, SCHMIDT *et al.* (*124*) have recently proposed a second explanation for the action of SOD. The studies of this latter group have led them to propose that NO^- rather than NO is the primary product of NOS turnover and that SOD mediates its oxidation to NO. This suggestion, if borne out by future studies, will require reformultion of the NOS-catalysed reaction.*

10.6.4. The H_4B Binding Site

Amino acids that contribute to the H_4B binding site of NOS have not been clearly identified, although sequence matching to other pterin-binding proteins has led to speculation about the location of this site. One region proposed (544) lies within the reductase domain and can, therefore, reasonably be excluded as proteolysis and expression studies (385, 391–395) unequivocally locate the H_4B binding site within the oxygenase domain of the enzyme.

A second region, residues 448–480 of murine iNOS (corresponding to 454–486 of the human isoform, Fig. 41), reportedly (355) exhibits a degree of homology with the H_4B-binding site of the aromatic amino acid hydroxylases. Subtle point mutations within this region (G450A and A453I) were indeed found (355) to attenuate H_4B binding without affecting the binding of heme. However, the same mutations also disrupted the dimeric structure of NOS and other studies (172, 523) have shown that disruption of dimer formation itself destabilises H_4B binding; thus G450 and A453 may not participate directly in H_4B binding, but rather contribute to the surface involved in subunit dimerisation. A more recent study by VENEMA et al. adds weight to the latter hypothesis (388). In this study mutation of the corresponding residues in dimeric H_4B-free bovine eNOS (G442A and A445I) substantially weakened the subunit interactions.

NISHIMURA and co-workers examined a third section of the enzyme (rat nNOS residues 558–721) which includes the region just discussed and is similar to the pterin-binding module of dihydrofolate reductase (DHFR) (510). However, this fragment of NOS, when expressed separately, failed to bind H_4B but instead bound the NOS inhibitor N^{ω}-nitro-L-arginine (NNA), suggesting that residues in this fragment contribute to the Arg binding site of NOS. Since the NISHIMURA study, mutational analyses by CHEN et al. (298) and STUEHR et al. (299) have pinpointed a residue within this region that specifically interacts with the substrate: E361 (human eNOS) and E371 (murine iNOS), homologous residues in the two proteins, directly bind the substrate guanidinium moiety. The corresponding residue in human DHFR (E30), according to

---

* See Addendum.

the Nishimura alignment with NOS, is known from crystal structure data (*299*, *545*) to interact with the cyclic guanidine function of the dihydrofolate pterin system. Therefore, the DHFR-homologous region of NOS clearly does not fulfil the same role as the pterin-binding module of DHFR which binds substrate and NADPH (*546*). The mutagenesis studies by Chen (*298*) and Stuehr (*299*) have shown that several other conserved acidic residues within the NOS DHFR module impact to varying degrees upon subunit dimerisation, heme content and H_4B / Arg binding. Earlier studies by Chen *et al.* (*381*, *382*) implicated C99 (human eNOS) in the modulation of H_4B binding to NOS; in particular a C99A mutation causes a reduction in H_4B binding affinity. Similar studies with iNOS and nNOS have shown that the corresponding Cys residue in these enzymes is also required for high-affinity pteridine binding (*387*). Thus, residues involved in H_4B binding by NOS are probably not restricted to the DHFR module but may straddle the heme binding site of the enzyme.

Boyhan and co-workers have also recently studied similar regions of the NOS oxygenase domain (*374*). Detailed sequence alignments between NOS isoforms and the aromatic amino acid hydroxylases revealed a region of modest similarity extending over 150 residues (475–625 in human nNOS). This sequence overlaps the DHFR-homologous region of NOS and was suggested to constitute the core region of the pterin-binding site in the enzyme; two sections of particular homology were identified within the region corresponding to human nNOS residues 475–510 and 589–624. However, expression and binding studies with truncated rat nNOS mutants also suggested that a critical determinant for Arg and H_4B binding is located within the region 275–350. It is noteworthy that this region includes the conserved Cys residue (C99 in human eNOS) identified by Chen *et al.* (*382*) as a key residue essential for H_4B binding.

Caution must be exercised in regards to the sequence alignment studies undertaken between NOS and other pterin-binding proteins. When comparing two or more proteins, the issue of prime importance is the homology of their 3D structures, since that is the best indicator of functional similarity. Comparison of the underlying amino acid sequence is an approximation and, with proteins that have had a long time to diverge, often a poor one. Clearly the substrate-binding glutamic acid residue located by Chen and Stuehr, and proposed to correspond to a pterin-binding residue in DHFR, does not fulfil the same function in the two enzymes.

Mayer and co-workers (*387*) have attempted to probe the function of the NOS DHFR-homologous module using synthetic peptides corresponding to subsections within the domain. Obviously this strategy requires that the peptides have sufficient structure in solution to mimic and compete with the endogenous sequences present in the full-length protein. Four sequences from rat nNOS were constructed, Fig. 55, and of these one, corresponding to residues 564–582 (A), exhibited concentration-dependent inhibition of Cit formation. This sequence did not interfere with H_4B or Arg binding to the enzyme nor did it inhibit electron transfer from the reductase domain to cytochrome c. However the peptide did inhibit the uncoupled reduction of O_2 to $O_2^{\cdot-}$ / H_2O_2 which occurs in the absence of Arg (*144, 345, 404*). The group concluded, therefore, that the 564–582 sequence of nNOS critically organises the tertiary structure of the enzyme for internal electron transfer from the reductase domain onto the heme unit. A 500- to 1000-fold excess of the synthetic peptide was required for half-maximal inhibition of the enzyme and, in view of this modest potency, it was suggested that position of the 564–582 sequence within the tertiary structure of the enzyme is not readily accessible from outside the protein. This idea was corroborated when an antibody raised against the synthetic peptide sequence failed to inhibit the enzyme, again indicating that the 564–582 region is buried within the enzyme.

That H_4B and Arg both bind in close proximity to the heme is evinced by the influence of each on the binding of the other (*501*), and by

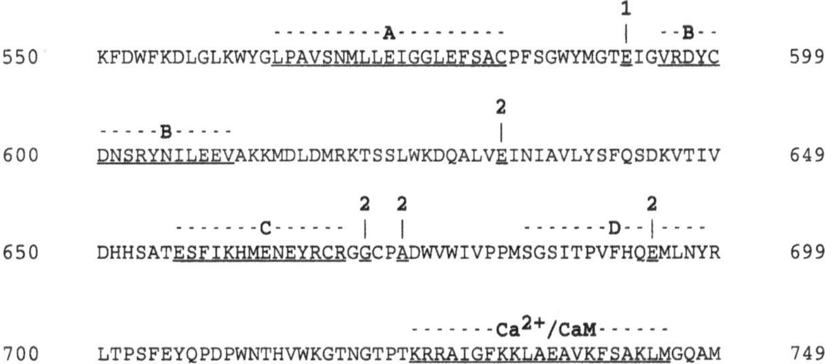

Fig. 55. Primary structure over the DHFR-homologous and CaM-binding zones of rat nNOS. MAYER *et al.* (*387*) constructed four synthetic peptides corresponding to regions **A-D** as marked. Only synthetic peptide **A** (corresponding to residues 564–582) inhibited the enzyme. Inhibition is assumed to occur by insertion of the synthetic peptide into the tertiary structure of the protein at the position normally occupied by the 564–582 sequence. This was shown to block electron transfer from the reductase domain of the enzyme onto the heme. Other key residues marked for reference purposes are as follows: **1**, glutamic acid residue that binds the substrate's guanidine moiety (*298, 299*); **2** residues required for subunit dimerisation (*299, 388*)

the influence of both on the heme spectral properties/spin state (*395, 481, 485, 486, 493, 547*) and binding of heme ligands (*406, 484, 488, 548*). Moreover, point mutation of the proximal Cys ligand, which prevents heme binding to the enzyme, also renders the enzyme incompetent with respect to H_4B and substrate binding (*384, 386*). Given the likely mechanism of NOS catalysis, involving reductive activation of O_2 at the heme centre, it is hardly surprising that the substrate should bind near to the heme unit. However, the mutual proximity of heme and H_4B is unusual and suggests a new role for the latter in enzyme mechanisms. This, as MAYER and WERNER have suggested (*173*), may reflect the unusual properties of the enzyme's reaction product, NO.

In summary, then, H_4B may function both as an allosteric effector and redox-active cofactor. H_4B binding to NOS stabilises the active dimeric state of the enzyme (*172, 353*). The cofactor also helps to maintain a stable heme pocket that is accessible to O_2 (*481*), and prevents or reverses inhibition of NOS by NO (*541*). Recent photoaffinity labelling studies with pteridine derivatives have confirmed that the H_4B-binding site is located in the region between the heme and CaM binding sites (*539, 549*). Further studies are required, however, to elucidate the detailed role of this cofactor in NOS catalysis and the complex biological chemistry of NO.

10.7. NOS Dimer Assembly

NOS isoforms share a common $\alpha2$ quaternary structure, derived from two identical monomeric subunits which consist of an N-terminal oxygenase domain and a C-terminal reductase domain. The enzymes are catalytically active only in the dimeric form (*145, 172, 353–357*) and, therefore, regulation of NOS subunit interactions could facilitate modulation of enzyme activity *in vivo*. Several studies with the inducible (*172, 355–357, 388, 392, 516, 550*), neuronal (*353, 374, 388, 523*) and endothelial (*388, 487, 515*) isoforms of NOS have examined the dependence of dimer assembly on substrate, cofactors and subunit domains. The availability of H_4B-free (*422, 428, 487, 489–492, 499*) and heme-free (*172, 523*) NOS preparations has been particularly useful in this work.

The determinants responsible for iNOS dimer formation are thought to residue entirely in the oxygenase half of the enzyme subunit (*392, 516, 550*), although its first 58 residues are not required for oligomerisation (*386*). Evidence for this is provided by the observation that the isolated oxygenase portion of the enzyme, but not the reductase domain, retains the capacity of the full-length enzyme for dimerisation (*392*). For this reason STUEHR and co-workers have proposed a head-to-head alignment model for the active dimeric enzyme (*392*). The importance of cofactors to the process of iNOS dimer assembly has been proven by mutagenesis of the heme-coordinating Cys residue (C200 in the human isoform) which destroys the enzyme's capacity for heme and H_4B binding and results in a monomeric protein (*386*). Other studies have also shown that dimerisation of the enzyme's subunits requires the coincident presence of H_4B and stoichiometric amounts of heme and Arg, although the latter does not remain bound to the enzyme following dimerisation (*172, 356, 518*). The process of dimerisation is somewhat slow, but proceeds with stoichiometric incorporation of heme and near-stoichiometric incorporation of H_4B (*172*). Dimerisation of dissociated heme-containing NOS monomers is also promoted by the presence of either Arg or H_4B, but optimally by the presence of both (*550*). Derivatives of Arg possessing a modified guanidine moiety have also been shown to promote dimerisation of iNOS, and their capacity to do so correlates, in the main, with their binding affinity for the dimeric protein (*493*). Interestingly, simple guanidine analogues such as thiourea and *N*-hydroxyguanidine also promote iNOS dimer formation, indicating that the α-amino acid component of Arg is not essential to the assembly process. Some evidence (Section 8.3) suggests that NO biosynthesis may be regulated by the availability of Arg, at least for this high-output isoform, but it remains to be seen whether Arg levels in intact cells limit dimer assembly.

A similar head-to-head assembly of nNOS and eNOS dimers is highly probable. Indeed structural analysis of the neuronal isoform has suggested that the dimeric protein exists in an elongated form (*353*). This is to be expected if the reductase domain consitutes a peripheral extension to the protein which exerts little influence over the oxygenase surface that participates in dimerisation. Moreover, BOYHAN *et al.* have recently found the isolated oxygenase domain of nNOS expressed in *E. coli* to exist predominantly as a dimeric protein (*374*). According to analysis by circular dichroism (*523*) the secondary structure of the heme-free nNOS monomer is very similar to that of the dimeric enzyme and, though incompetent for NO synthesis, the protein retains the functional reductase activity of the dimeric enzyme, as evidenced by its ability to reduce external oxidants such as cytochrome *c* (*172, 523*). The heme-free subunit, however, lacks the capacity for stable incorporation of H_4B and is unable to bind substrate analogues such as NNA (*518, 523*). In contrast, dissociation of H_4B from dimeric heme-containing nNOS is very slow (*523*). Other work with the neuronal isoform (*353*) suggests that the presence of bound H_4B strengthens the subunit interaction within the dimeric enzyme and confers resistance to dissociation induced by sodium dodecyl sulfate (SDS).

Dissociation of dimeric iNOS is accompanied by full conversion of heme iron into its low-spin configuration due to the presumed binding of water or hydroxide as a sixth ligand (*550, 551*). Subunit dissociation also increases accessibility to bulky heme-coordinating ligands such as DTT (*550, 551*). These observations suggest that dissociation exposes the distal heme pocket to solvent in a manner consistent with the heme occupying a position at the subunit interface, where it might participate in a single catalytic site (*358, 374*). However, *dimeric* H_4B-free NOSs expressed in bacteria also contain fully low-spin heme and can bind DTT as a sixth ligand. Thus the increase in solvent accessibility of the heme centre could well be due to the loss of H_4B which occurs on dissociation rather than the heme necessarily occupying a position at the subunit interface. Moreover, other evidence (see Section 10.5) has been interpreted in favour of a dimeric structure possessing two catalytic sites that are structurally linked but capable of independent function.

Contrasting observations in a number of studies suggest that the NOS isoforms may differ from one another in terms of their requirements for dimerisation. For example, heme-free nNOS subunits, unlike those of the inducible isoform, dimerise in the presence of heme alone and do not require H_4B (*523*). Work with eNOS (*487*) similarly fails to support a role for H_4B in promoting the formation of the dimeric enzyme, although in this instance the presence of H_4B, which modifies the heme environment, was found to stabilise the enzyme towards heme loss. The endothelial isoform also differs from the other two in that dimer formation reportedly involves determinants in both oxygenase and reductase domains (*515*).

Dimer dissociation and subunit unfolding have been studied with urea as a denaturant. In the case of iNOS these two processes occur as separate transitions within two distinct urea concentration ranges (*551*). At urea concentrations of up to 2.5 M dissociation of the dimeric protein proceeds with the release of bound H_4B to liberate increasing amounts of the inactive monomeric subunit in which the sixth coordination site of the heme iron is exposed to solvent. Dissociation is reversible, however, in that the catalytically active dimeric enzyme can be reconstituted in the presence of H_4B and Arg. The monomeric subunit retains the other cofactors (Cys thiolate-ligated heme, flavins and CaM) as well as the protein folding because full reductase activity is maintained. Further studies (*516*) have shown that dissociation disables electron transfer from the reductase domain to the heme moiety, possibly because the heme iron converts fully to a low-spin state in the monomeric protein.

This is significant and suggests that dimerisation may control NOS activity by modulating interdomain electron transfer within the protein. When iNOS is exposed to urea concentrations above 2.5 M partial unfolding of the monomer ensues; although the bound flavins and CaM are retained, heme is released and the reductase activity is lost (*551*). The unfolding of iNOS, unlike dimer dissociation, is only partially reversible. Dissociation of dimeric nNOS requires higher concentrations of urea than the levels that dissociate the inducible isoform, suggesting a weaker dimeric interaction for the latter (*396*). Interestingly, higher urea concentrations are also required to dissociate a truncated iNOS oxygenase domain dimer compared to the full-length iNOS dimer; thus the attached reductase domains may destabilise the dimeric interaction between oxygenase domains in this isoform (*550*).

VENEMA *et al.* have recently expressed the three mammalian isoforms of NOS in the same experimental systems in order to allow direct comparative analysis of factors that contribute to dimerisation (*388*). The endothelial isoform expressed in an insect cell line or in endothelial cells was found entirely in the dimeric form, unlike earlier studies (*487*) with a prokaryotic system (*E. coli*) in which overexpression of the enzyme afforded a monomer-dimer mixture. The dimerisation of eNOS was found to require incorporation of the heme prosthetic group, but was independent of myristoylation or the binding of CaM and H_4B (*388*).* However, incubation with the latter cofactor was observed to stabilise the

---

* Expression of eNOS in insect cells by the VENEMA group was carried out in the presence of 2,4-diamino-6-hydroxypyrimidine (DAHP) – an inhibitor of H_4B biosynthesis (*388*). The observation that H_4B-free eNOS exists entirely in a dimeric state demonstrates that this cofactor is not required for subunit association. However, another recent study by MAYER *et al.* (*519*) has found that expression of eNOS in a very similar system produces a mixture of monomeric and dimeric protein. TÓTH *et al.* (*552*) have also recently found H_4B to promote dimerisation of monomeric eNOS subunits in placental microsomes. These discrepancies require clarification and may possibly arise from differences in the expression conditions or purification procedures used to isolate the enzyme. Although it is clear that H_4B consistently stabilises the eNOS dimer against dissociation induced by SDS, VENEMA'S results strongly suggest that H_4B is not absolutely required in order to promote eNOS subunit association. In keeping with this, HELLERMANN and SOLOMONSON have also recently found (*553*) dimerisation of a truncated eNOS protein comprising the oxygenase domain and CaM-binding site, but lacking the reductase domain, to be independent of Arg and H_4B. However, these authors demonstrated that self-association did not proceed with a truncated protein that lacked the CaM-binding sequence as well as the reductase domain and concluded, therefore, that CaM *does* promote the dimerisation of eNOS. HELLERMANN and SOLOMONSON speculate that for eNOS dimerisation may be a primary event in the activation of the enzyme by Ca^{2+}. Other reports reveal that the

dimeric enzyme to dissociation induced by SDS. The presence of Arg has little influence over the stability of the dimer. For this isoform, then, it is unlikely that H_4B- or substrate-induced dimer formation contributes to the regulation of eNOS activity in endothelial cells. The VENEMA group found expression of nNOS to afford a 6:4 monomer-dimer mixture and, in contrast to eNOS, the presence of either Arg or H_4B, but optimally both factors, increased the extent of dimerisation as well as the stability of the dimer to dissociation. Although H_4B promotes dimerisation of the neuronal enzyme *in vitro*, these studies confirmed that the cofactor is not absolutely required for dimerisation of this isoform. The same work found the iNOS dimer to be the least stable of the three isoforms with respect to dissociation and, when expressed, this enzyme was found mainly in a monomeric state consistent with earlier studies (*172, 357*). The absolute dependence of iNOS oligomerisation on H_4B suggests that the quaternary structure, and thus the activity, of this isoform is more likely to be sensitive to regulation by H_4B than the two constitutive NOS isoforms.

VENEMA *et al.* (*388*) also examined domain interactions that contribute to subunit association and found that the determinants for iNOS dimerisation reside exclusively in the oxygenase domain, thus confirming the head-to-head dimerisation model. However, association of eNOS and nNOS subunits within their respective dimeric enzymes was shown to involve not only head-to-head interactions, but also head-to-tail and (more weakly) tail-to-tail interactions between the subunits. The more extensive subunit interactions in the constitutive isoforms likely accounts for their greater stability in the dimeric state. Sections within the oxygenase domain that contribute to the dimer interface have

---

participation of CaM in the overall regulation of eNOS activity is subtle. Thus, STUEHR'S group have shown (*313*) that eNOS has a high affinity for this regulatory protein and that even under resting conditions Ca^{2+} levels are sufficient to ensure a basal activity for the enzyme; the activity of the eNOS is, therefore, subject to other mechanisms of control. Meanwhile MICHEL'S group have found (*471*) that Ca^{2+}/CaM binding to eNOS also serves to disrupt an inhibitory heteromeric complex between the enzyme and caveolin, a structural protein of the caveolar regions to which eNOS is targeted in endothelial cells. As CaM binding appears to facilitate dimer assembly in eNOS it is curious that the Ca^{2+}–independent, inducible isoform, which in contrast to eNOS contains essentially permanently bound CaM, is expressed mainly in a monomeric state. This underlines the subtle differences between the NOS isozymes which undoubtedly reflects the differences in the cellular environments in which they operate and the various physiological roles for which they produce NO. The greater H_4B-dependence of iNOS dimerisation suggests, perhaps, that control over levels of this cofactor are important for the regulation of iNOS activity. It is significant, therefore, that in stimulated immune cells H_4B biosynthesis is coinduced with iNOS (see Section 10.6).

yet to be identified in detail. However, mutation of two residues which are conserved in all known NOS sequences, G450 and A453 in murine iNOS, is known to abolish both dimerisation and H_4B binding in the inducible isoform (*355*). These residues occur in a sequence (448–480) which possesses homology to three different H_4B-utilising aromatic amino acid hydroxylases (*374*). In view of the interdependence of H_4B binding and subunit association for iNOS, it was unclear whether these residues contribute to an H_4B-binding site or to the dimerisation interface of the protein. VENEMA *et al.* (*388*) have clarified this issue by mutational analysis of the corresponding residues in bovine eNOS (G442A and A445I mutations). A significant reduction in the strength of the dimeric interaction between eNOS oxygenase domains was observed in this study. Since dimerisation of eNOS is independent of added H_4B, it is clear that G442 and A445 affect dimerisation directly rather than indirectly by loss of H_4B binding. These residues, then, are likely to mark an important section of the protein that contributes to the subunit interface involved in dimerisation. Another recent study by STUEHR *et al.* (*299*) examined the role of several conserved acidic residues in the oxygenase domain of murine iNOS by alanine screening mutagenesis. Mutation of residues E411 and E473 (corresponding to E401 and E463 in human iNOS, Fig. 41) destabilised the dimeric structure of iNOS and afforded predominantly monomeric protein. However, in contrast to the E411A mutant, which exhibited normal H_4B and Arg binding characteristics, analysis of the E473A mutant suggested a loss of binding affinity for H_4B and Arg. Curiously, several other mutants which were to some degree defective in either H_4B and/or Arg binding retained a dimeric structure. Thus, the absolute dependence of iNOS dimerisation on Arg and H_4B has been called into question, at least in the case of its isolated oxygenase domain overexpressed in *E. coli*.

In summary, it is clear that when catalytically active NOS enzymes are dimeric. The presence of heme is essential for dimerisation of each of the enzymes, but the importance of H_4B to subunit association varies from one to another. Despite this H_4B is an essential cofactor required for NO biosynthesis regardless of isoform. Thus, the role of H_4B in NOS catalysis involves more than promotion of oligomerisation and stabilisation of the dimeric enzyme.

10.8. Autoinactivation/NO Feedback Inhibition of NOS

NO has been shown to inactivate NOS through ligation to the heme iron centre (*136, 484, 541, 554–562*) in a manner analogous to its binding to many heme proteins such as guanylate cyclase and hemoglobin (*563, 564–568*). It has emerged in recent years that NO can bind both the ferric and ferrous forms of NOS heme, although NO binds to the reduced form more strongly than it does to the oxidised form (*484*). The ferrous-nitrosyl complex, though stable under anaerobic conditions (*484*), decomposes in the presence of O_2 to regenerate ferric heme (*562*). NO may, therefore, constitute a feedback inhibitor which regulates its own production through binding to NOS heme, but the instability of the heme nitrosyl complexes to O_2 has caused some to question the relevance of this mechanism to regulation of the enzyme *in vivo* (*560*). However, Stuehr and co-workers have suggested that 70–90% of nNOS exists as the inactive ferrous-NO complex during steady state NO production (*561, 562*). Moreover, nitrosyl complex formation was found to be independent of added NO scavengers, indicating that heme complexation occurs with NO that has not left the enzyme's catalytic site. Stuehr proposed that it is the O_2-dependent decay of the ferrous-NO complex, rather than the intrinsic affinity of heme iron for O_2, that is the limiting factor which governs NOS activity by controlling the amount of free NOS circulating. Thus, in addition to the regulatory mechanisms mentioned in section 8.3 it appears that NOS activity, at least for the neuronal isoform, is coupled to the physiological O_2 concentration in a given tissue. It remains to be seen whether these insights lead to the disclosure of new aspects of the complicated regulatory role of NO in living organisms.

It is not yet clear how structural aspects of the NOS catalytic site may influence the susceptibility of NOS heme to complexation by NO. Arg binding to NOS reduces the affinity of ferric heme for NO, but ferrous heme retains its affinity and produces a more stable nitrosyl complex when Arg is bound to the enzyme (*484, 560*). Stuehr and Griffith (*119*) have suggested a number of possible mechanisms (that have not been substantiated) through which the heme susceptibility to complexation by NO might be moderated:

(i) According to the proposed mechanism for NOS monooxygenation II, ferric heme (Fig. 30) is initially regenerated at the end of the cycle with an axial hydroxide ligand. Inactivation of the heme would be reduced if the rate of NO diffusion out of the catalytic centre is more rapid than the loss of water from the heme iron.

(ii) The Cit co-product, if resident at the catalytic site for a longer period than NO, may possibly screen the iron centre from attack by NO.

(iii) An axial ligand contributed by an active site amino acid residue could block the distal site at the end of the catalytic cycle.

Other mechanisms by which NO modulates the activity of NOS have also been postulated. PATEL *et al.* (*569, 570*) have proposed that NO mediates the oxidative coupling of vicinal thiols in the enzyme to form disulfide bridges that regulate enzyme activity by causing a change in its conformation. For the inducible isoform a recent study of murine macrophages has suggested that NO synthesis can also limit dimer assembly and, therefore, regulate the enzyme's activity (*357*). NO synthesis is known to inhibit heme production, but in these studies even the addition of heme precursors or hemin itself failed to increase heme insertion into the enzyme or to promote dimer assembly in the cells generating NO. Thus it is likely that NO antagonises heme insertion into the NOS apoprotein. Other work suggests that NO may regulate its own production by interfering with the interaction of the NOS gene promoter region with NF-κB, a DNA binding protein that takes part in the induction of iNOS gene transcription; this modulation may also serve a negative feedback function to limit production of NO *in vivo* (*571*). NOS gene expression and expressional control have been reviewed elsewhere (*572, 573*) as has the more general regulation of NO biosynthesis (*197*).

NOS is generally regarded as a moderately unstable enzyme to work with and it is also possible that irreversible inactivation of the enzyme may occur by reaction of essential cofactors and protein residues with ROS. These species may, of course, be generated by NOS itself when O_2 reduction is uncoupled from substrate turnover (*144, 345, 404, 540, 574, 575*), but their formation, which may only be relevant to the neuronal isoform *in vivo* (*519*), is minimised in the presence of sufficient concentrations of Arg and H_4B (*345, 540*). Additionally the reaction of NO itself with superoxide generates highly reactive peroxynitrite which may inactivate NOS. Indeed recent work suggests that peroxynitrite causes irreversible destruction of the heme Cys thiolate ligand or otherwise perturbs heme or its environment (*576–577*). This is consistent with observations that thiols such as 2-mercaptoethanol stabilise the enzyme by preventing oxidation of essential protein Cys residues (*517, 578, 579*).

10.9. Distinctive Features of NOS Isoforms

Considerable variation in their specific activities has been reported for NO synthases isolated from different sources, but these differences are likely to reflect variable cofactor loss in isolation procedures, enzyme instability and differences in activity assays. The highest specific activities recorded for the three isoforms, however, all fall within a relatively narrow range, and it has been argued that this similarity reflects an identical chemical mechanism and comparable catalytic efficiency operating among the isoforms (*119*). However, only modest sequence identity (50–60%) exists between different isoforms, although sequences are highly conserved for any single isoform across different species (*143*). Thus, each of the constitutive NOS isoforms is 90–95% identical across mammalian species and the amino acid sequences of human and murine iNOS enzymes share about 80% identity (*163*). Despite the modest sequence identity between different isoforms, they all exhibit a high degree of homology in the protein regions associated with cofactor binding sites and this, together with inhibitor studies, has also been taken as an indication that the isoforms share similar if not identical catalytic mechanisms for the conversion of Arg into Cit and NO (*326*). The isoforms are distinguished from one another by their sub-cellular targeting and regulation, although constitutive expression of eNOS and nNOS is frequently used to group these two isoforms together as distinct from the inducible enzyme. It has been suggested, however, that the inducible isoform may also effectively be expressed constitutively in certain tissues and, therefore, that this enzyme is probably best understood as the isoform which is independent of Ca^{2+} levels (*197, 261*).

A recent report from Stuehr's (*313*) laboratory indicates that the constitutive isoforms are more divergent in their properties than has previously been thought to be the case. Indeed eNOS has characteristics that are intermediate and, in some cases, completely distinct from nNOS and iNOS. Firstly, the activity of eNOS with regards to NO synthesis was found to be significantly lower than either the neuronal or inducible isoforms, and a slow electron transfer step between FAD and FMN was postulated to account for this. Another distinction between the two constitutive isoforms was seen in the effect of substrate on NADPH oxidation rate. The substrate-uncoupled oxidation of NADPH by eNOS, in the presence of Ca^{2+}/CaM, was 13 times lower than that of nNOS, and much more comparable to the rate of uncoupled NADPH oxidation by iNOS (*183*). The fact that uncoupled NADPH oxidation by iNOS is slower than by nNOS is surprising in view of the comparable specific activities for NO synthesis by these two isoforms. This observation has prompted Griffith and Stuehr to suggest that uncoupled oxygen reduction by iNOS is down-regulated in order to prevent continuous generation of ROS by a NOS isoform that may have to operate in an Arg depleted environment (*e.g.* wounds) (*119*). In the presence of Arg the rate of NADPH oxidation is increased two-fold and six-fold for eNOS and iNOS respectively; the NADPH oxidase activity of the neuronal isoform, however, is actually slightly reduced in comparison to its rate of uncoupled NADPH oxidation (*313*). Therefore, CaM binding alone is sufficient to promote maximal electron transfer and heme reduction in nNOS. On the other hand eNOS requires binding of substrate as well as CaM for full activation of the enzyme's internal electron transfer pathway. For the Ca^{2+}–independent inducible isoform regulation of heme reduction by the substrate is even more pronounced.

In these studies STUEHR'S group also demonstrated that eNOS has a higher binding affinity for CaM than has generally been recognised. Indeed the enzyme co-purified with CaM, although studies with a Ca^{2+} sequestering agent did demonstrate that CaM binding to the enzyme is reversible. In retrospect the high affinity of eNOS for CaM is not surprising as VENEMA *et al.* have shown (*414*) that the isolated eNOS CaM recognition sequence binds the protein almost as effectively as the corresponding sequence from iNOS. The high affinity for CaM means that in stimulated endothelial cells Ca^{2+} ion levels are sufficient for maximal activation of the enzyme, but even under resting conditions Ca^{2+} ion concentrations are such that eNOS always has a basal activity. High affinity for CaM may ensure a moderate and consistent production of NO in the cardiovascular system without susceptibility to transient temporal variations in Ca^{2+} ion concentrations; this finding may explain why the enzyme has evolved with a sensitivity to regulation by the substrate. In contrast to eNOS, the neuronal isoform exhibits half-maximal activation at significantly higher Ca^{2+} ion concentrations. This results in an activity profile which is subject to greater regulation by elevation in intracellular Ca^{2+} levels and may explain why the enzyme's activity does not require modulation by substrate binding.

The emerging picture of NOS, then, is one of an elongated, head-to-head, symmetrical dimer with each subunit organised into two distinct functional domains, a reductase domain and an oxygenase domain. The former facilitates electron transfer from NADPH *via* tightly bound flavin cofactors to the catalytic site located in the oxygenase domain. Residues at the dimer interface are contributed exclusively, or at least to a very large degree, by the oxygenase domain. Heme, H_4B and Arg are bound in close proximity at the catalytic centre. The weight of evidence currently favours a dual-site model, that is, one in which there are potentially two functioning catalytic sites (one in each subunit) as opposed to a model in which a single catalytic site is formed jointly between the two subunit oxygenase domains. Evidence suggests that the structure of each catalytic site is reciprocally maintained by the presence of bound heme in the other subunit. The binding of Arg and H_4B at the same catalytic site exhibits positive cooperativity. In contrast, binding of a molecule of H_4B at both catalytic centres is subject to pronounced anticooperativity, leading to the tight binding of this cofactor in the catalytic centre of just one of the two subunits. For this reason it is thought that NOS actually operates with just one of the two catalytic centres producing NO at any given moment. The two functional domains of each subunit are connected by a short regulatory, CaM-binding sequence. CaM binding may control NOS activity both by enhancing intradomain electron transfer within the reductase domain and by facilitating interdomain electron transfer from FMN onto heme. It is clear from slight differences in the interactions with CaM, other cofactors (heme/H_4B) and substrate that each isoform of this uniquely complicated class of enzymes is distinct from the others. Each is finely tuned to the biological environment in which it operates, and each is elegantly regulated by a subtle combination of factors suited to the generation of NO for different physiological purposes in different tissue types.

11. Clinical Uses of NO and Its Inhibitors

An overview of the clinical uses of NO and its inhibitors cannot hope to encompass all the areas in which NO is implicated as new reports appear almost daily. Instead it will concentrate on the more well developed aspects of NO biology, and thus whet the appetite of the reader.

11.1. Respiratory Systems

NO is released from pulmonary arterial and venous endothelial cells in response to agonists such as acetylcholine, and some work has suggested that patients with primary pulmonary hypertension have a basic defect in the endothelial arginine-NO pathway (*580*). Inhaled NO has been shown to reduce pulmonary constriction without causing general hypotension in healthy volunteers (*581*). The authors speculated that the inhaled NO is rapidly inactivated to form nitrosylhemoglobin, which is then oxidised to methemoglobin thus restricting vasodilation in the pulmonary vasculature and preventing systemic vasodilation. Using animal models, several groups have also shown improvements in experimentally induced pulmonary hypertension with inhaled NO (*582, 583*), though this has not been confirmed by others who have found only a partial reversal of pulmonary vasoconstriction (*584, 585*).

The use of inhaled NO has been investigated extensively in newly born babies where pulmonary hypertension is a major cause of morbidity and mortality. ROBERTS *et al.* (*586*) have shown that there is an improvement in pre and post-ductal oxygen saturation in children with Persistent Pulmonary Hypertension of the Newborn (PPHN) treated with inhaled NO and selective pulmonary dilation in children undergoing cardiac catheterisation given inhaled NO. Several studies have shown benefit from low doses of inhaled NO (*581, 587*) and one group investigating the dose response relationship of inhaled NO found no significant difference between high and low concentrations (*588*). The optimum dosage of inhaled NO in this group of patients has yet to be formalised, but the therapy has been widely adopted in a neonatal setting and the dose is currently estimated on an individual response basis. Endothelial dysfunction may play a central role in pulmonary hypertension, and the vascular response to acetylcholine is altered in a number of conditions associated with pulmonary hypertension where NO production is inadequate. Intraoperative and postoperative uses of inhaled NO have also been described in conditions associated with pulmonary hypertension, including mitral valve replacement with pulmonary hypertension (*589*) and postoperative complications after operations in children and young adults with congenital heart diseases (*590*). Increased pulmonary vascular resistance is commonly seen during the postoperative care of cardiac surgical patients, and this increased afterload on the right ventricle may worsen cardiac output by compromising right ventricular function. Vasodilators currently used act as both systemic and pulmonary vasodilators and can result in profound systemic hypotension (*591*). Cardiopulmonary bypass itself has been implicated in the endothelial injury leading to increased pulmonary vascular resistance. WESSEL *et al.* (*592*) compared the effectiveness of acetylcholine and NO in reducing pulmonary artery pressure after cardiac bypass and showed that cGMP was significantly higher when NO was inhaled, but unchanged when an acetylcholine infusion was given, from which they inferred endothelial dysfunction. Many of the data in humans have focused on patients with both pulmonary hypertension and hypoxemia, and since hypoxemia is an important cause of pulmonary vasoconstriction, the effect of NO on pulmonary vasoconstriction itself has not been well established (*593*).

Inhaled NO has been shown to be of benefit in severe allograft dysfunction (*i.e.* acute rejection) of transplanted human lungs. Acute severe rejection develops in 10–20% of lung transplant recipients (*594*) and is thought to be due to injury resulting from re-entry of oxygen into tissue which had previously been starved of oxygen (*595*). TRIANTAFILLOU *et al.* (*596*) examined the effects of short periods of NO inhalation after bilateral lung transplantation in humans. In patients with hypoxemia, NO lowered pulmonary artery pressure with no systemic effects, but interestingly worsened oxygenation in patients with normoxia. The same group have subsequently shown improved oxygenation in a larger group of patients with allograft dysfunction. However, these results must be interpreted with caution as other evidence implicates NO in the production of allograft rejection, where a NOS inhibitor delayed the onset and decreased the severity of rejection in a rat model (*597*). The same group have also shown attenuation of established cardiac allograft rejection with NOS inhibition, also in a rat model (*598*). A recently published article on human cardiac allografts has shown an association between induction of NOS and left ventricular dysfunction (*599*). It is likely that, as in inflammation and sepsis, NO in allograft dysfunction has both a protective role to play as well as a destructive one with the net result dependent on the balance of the two processes.

11.2. Airway Modulation and Asthma

Asthma is an inflammatory disease with a recoginsed yearly mortality rate. Those who die from acute asthma have grossly inflamed airways with mucous plugging (*600*). It has recently been shown that exhaled air in asthmatics and patients with other inflammatory conditions such as bronchiectasis and lower respiratory tract infection contains higher concentrations of NO than controls (*601*). The source of this NO appears to be from the lower airways (*602*). L-NMMA given to normal subjects by inhalation causes a decrease in exhaled NO but no systemic effects (*603*), suggesting it is localised to the respiratory tract. Non-selective NOS inhibitors, such as L-NAME, cause decrease in exhaled NO in both asthmatics and controls wheareas aminoguanidine decreases exhaled NO in asthmatics only, implying that cNOS is the main source of NO in normal whilst iNOS is important in asthmatic airways (*604*). Whilst it is largely agreed that exhaled NO is a marker of airways inflammation there is controversy over the role of NO as there are both beneficial and deleterious effects. NO mediates NANC neural inhibitory responses in human airways and inhaled NO acts as a bronchodilator (*605*). It is not however an extremely effective bronchodilator, and whilst NOS inhibition in animals increases bronchoconstriction this is not seen in patients with asthma (*603*). The inflammatory response to infection is beneficial but, in the context of asthma, it is likely to increase the symptoms of airways obstruction. Whilst NO donors relax human airways *in vitro* if given systemically the dose is limited by cardiovascular side effects because of the greater effect of these compounds on vascular smooth muscle than airway smooth muscle (*606*).

11.3. Toxicology

There are potential hazards with the use of inhaled NO. At higher concentrations it can produce methemoglobinemia, although this is not a widespread problem since the concentration of inhaled NO used are low. However, there are reports of transient methemoglobinemia (*607*). There are also reports of pulmonary edema formation with inhaled NO (*608, 609*) which is likely to be due to acute improvement in right ventricular function in the face of a failing left ventricle. It appears that of inhaled NO 70–80% forms nitrate, but some nitrite must also be formed. Nitrite administration is associated with pulmonary edema which accounts mainly for its lethality. It is highly unlikely, however, that the one hour LD50 dose of NO_2 could be achieved (174 ppm) (*610*) given the doses of NO administered (10–80 ppm). There is a need to define the lowest effective concentrations of NO for each indication, and to provide safe guidelines for its administration. Inhaled NO is currently not formally approved though its use is widespread. All samples of NO will be contaminated with some NO_2. There are no standardised cylinders for NO delivery, and no standard administration systems to deliver NO to breathing circuits that are carefully tested for NO_2 production. Most systems used currently are for use in ventilated patients, and thus a standard administration for ambulant patients is also required.

11.4. Adult Respiratory Distress Syndrome

The lungs respond to a number of different pathologies *via* common pathways, producing diffuse damage. There is a spectrum of disease ranging from mild lung injury to Adult Respiratory Distress Syndrome (ARDS). The etiological factors fall largely into two groups: those arising from within the lung such as pneumonitis, pneumonia and lung trauma and those arising from without such as sepsis, trauma, burns, pancreatitis and systemic hypoperfusion. Alveolar collapse within the lung causes mismatch of the ventilated and perfused areas of the lungs (V/Q mismatch). Those areas which are perfused but not ventilated respond by locally constricting their blood vessels, thus diverting blood flow to better ventilated areas. Increased local production of NO early in ARDS may increase V / Q mismatching by causing vasodilation of blood vessels to non-ventilated parts of the lung (*611*). Established ARDS has a mortality rate of nearly 60%, and those with severe pulmonary hypertension have a higher mortality, though survivors have significant recovery of their lung function (*612*). Therefore there is considerable benefit to improving mortality and, in particular, to concentrate on the reduction of pulmonary hypertension.

Inhaled NO has been shown to reduce the pulmonary hypertension associated with ARDS (*613*). In low concentrations (2–40 ppm), inhaled NO produces selective vasodilation of pulmonary vessels in well ventilated areas of the lung without affecting poorly ventilated areas, thus improving V / Q matching (*614, 615*). This decrease in pulmonary pressure is thought to result in improvement in right ventricular function (*616*). Rossaint *et al.* (*617*) first published an article reporting beneficial effects of inhaled NO on mean pulmonary artery pressure (MPAP) and pulmonary gas exchange in patients with severe ARDS using concentrations of 18 and 36 ppm NO. A recent study using a canine model has suggested that NO is released from lung tissue during injury and that administration of an inhibitor increases the MPAP (*618*). Further work by Kavanagh *et al.* (*619*) has shown protection against increased PAP with inhaled NO. This further confirms the role of NO in control of pulmonary vascular tone which is disordered in acute lung injury. Rossaint *et al.* (*620*), in a retrospective study of 87 patients, showed improved arterial oxygenation and reduced PAP in most of those treated with inhaled NO, but no improvement in outcome, which may represent the fact that the most common cause of death in ARDS is multi-system organ failure and sepsis (*621*). Indeed, it seems that some patients can be termed "responders" to NO and others "non-responders". Although studies to date have not shown a significant improvement in the outcome of patients with ARDS treated with inhaled NO, results from a large prospective trial in North America are awaited with interest. Inhaled NO may, by virtue of decreasing PAP and improving oxygenation, "buy time" for other treatment modalities in ARDS to take their course of action.

11.5. Sepsis

Septic circulatory shock (sepsis) is characterised by hypotension, vascular injury and disseminated intravascular coagulation. The most common cause is infection with Gram-negative bacteria, leading to the release of "endotoxin", the lipopolysaccharide (LPS) comprising the outer membrane of the bacterial wall. LPS causes septic shock through the complicated activation and release of humoral mediators and cytokines. The hypotensive element to septic shock, which is characteristically unresponsive to vasopressor elements, has been poorly understood and yet it is the variable which, because it cannot be adequately controlled pharmacologically in the ITU setting, often causes irretrievable cardiovascular decompensation and death. Overall estimates suggest that 40% of septic patients develop septic shock of these mortality may be as high as 77–90% (*622*).

LPS and other mediators of sepsis stimulate iNOS gene transcription and translation, principally in macrophages (*623, 624*). This augmentation of iNOS expression resulting in excess production of NO is implicated in the hyporesponsiveness to vasopressors, and accumulating evidence indicates that it can account at least in part for the hypotension seen in sepsis. However, to date, there is little convincing evidence of increased NO output in human macrophages following iNOS induction, though several groups have demonstrated iNOS mRNA expression (*625, 626*).

A number of studies in humans have supported the hypothesis of increased NO production in sepsis (*627–629*). Septic shock is associated with increased production of urinary nitrate and nitrite, stable end-products of nitric oxide metabolism. These observations have led to the suggestion that NOS inhibitors might be useful in the treatment of circulatory shock. However, the situation is more complicated. The positive benefit of NO release as a cellular protective mechanism and certain deleterious effects of NOS inhibition systemically (*630*) have also been demonstrated. Since NO also inhibits microthrombosis formation by reducing platelet adhesiveness, and maintains vital organ perfusion by acting as a vasodilator in a condition where vasoactive mechanisms are also activated, inhibiting its production is clearly akin to walking a tightrope and calls for careful control. Many studies of sepsis in rat models have shown an early activation of eNOS, followed by a delayed and more prominent induction of vascular iNOS in response to systemic LPS (*631*). This leads to early non-sustained hypotension followed by a secondary hypotension which is unresponsive to vasopressors. Both responses are inhibited by NOS inhibition but only the delayed response is prevented by dexamethasone, thus strongly implicating iNOS in the sustained hypotension seen in shock (*632, 633*). iNOS knockout mice have, to some extent, resolved the role of NO in LPS-induced hypotension, although there remain conflicting data. iNOS deficient mice do not develop increased serum nitrite and nitrate levels when challenged with endotoxin (*634, 635*). Whilst Wei *et al.* (*634*) have shown that iNOS mutant mice are more resistent to endotoxin challenge, other studies have not supported this: Laubach *et al.* (*635*) showed no protection from the lethal effects of LPS in iNOS deficient mice. Also MacMicking *et al.* (*636*) showed an important protective effect of iNOS against fatality following *Listeria monocytogenes* infection.

11.6. NO, Inflammation, and the Immune System

L-Arginine can be metabolised not only to form NO and citrulline with NOS but also, with arginase (another LPS-inducible enzyme), to produce urea and L-ornithine. Ornithine is the substrate for the production of polyamines and L-proline, both of which are essential for tissue repair (*637*). In wound repair there is an early phase of iNOS activity followed by prolonged arginase activity (*638*). On the other hand, NO can cause inflammation. Thus inflammation and wound healing are a balance between pro-inflammatory and anti-inflammatory mechanisms and are normally self-terminating. NO produced by iNOS is implicated in the pathogenesis of inflammation, but there is also evidence of its anti-inflammatory properties as discussed previously with reference to the respiratory system. Pharmaceutical companies, who are currently developing specific iNOS inhibitors are therefore faced with the additional task of producing tissue-specific or locally acting iNOS inhibitors in order to limit unwanted side-effects. Extrapolating from the above observations, it may be that sepsis represents persistent uncontrolled overexpression of proinflammatory cytokines and iNOS, whereas certain fibrotic diseases such as fibrosis following ARDS may represent persistent uncontrolled wound repair (*637*).

As more roles of NO are discovered, so its role in sepsis becomes less straightforward, with both beneficial and detrimental effects well described. It is likely that the protective mechanisms initiated in the body become uncontrolled but that NO production by eNOS, and even during early iNOS induction, is of benefit by virtue of maintaining organ perfusion, inhibiting microthrombosis (especially in the mircrocirculation of vital organs), protecting the endothelium and acting as an anti-microbial and anti-inflammatory agent. It is probable that the conflicting data resulting from blanket inhibition of NOS in sepsis results from the reversal of these positive benefits. It is also important to note that there are more vasoactive substances involved in the regulation of hemodynamic control in sepsis than NO, *e.g.* endothelin-1, and that inhibition of NO may not simply reverse excess vasodilation but unmask and amplify vasoconstriction, particularly in susceptible organ beds such as the renal and gut circulations.

11.7. NOS Inhibition in Sepsis

Despite the above cautions, there are accumulating data for the effects of NOS inhibitors in septic shock. There are several levels at which NO production can be inhibited. Glucocorticosteroids (GCS) block the expression of iNOS, and pretreatment with steroids can prevent the induction of iNOS and vascular collapse seen in rats with endotoxic shock (*632*). However GCS do not alter the course of sepsis when given after LPS (*639*), and similarly have shown no benefits in trials in human septic shock (*640, 641*), principally because by the time of administration, iNOS is already induced. Inhibition of NOS can also be achieved by the use of L-arginine analogues, depleting enzyme cofactors or limiting substrate availability. The most widely researched of these is the use of L-arginine analogues. Studies of NOS inhibitors have given somewhat conflicting results. HOLLENBERG *et al.* (*642*) have shown reversal of the hyporesponsiveness to vasoconstrictors in an animal model of sepsis using videomicroscopy and other studies have shown improved outcome in rodent models of sepsis (*643*). Others however have shown that L-NMMA does not significantly improve outcome in sepsis in animal models (*644, 645*). There are also marked differences between animal and human models, which limit the extrapolation of data from these sources. However studies in humans have not proved more conclusive: PETROS *et al.* (*646*) first reported that NOS inhibition with L-NAME of L-NMMA caused an increase in MAP. Subsequently the same group showed that an infusion of L-NMMA produced an increase in systemic vascular resistance, pulmonary vascular resistance, and decreased cardiac output and heart rate (*27*). Subsequent reports of the use of NOS inhibitors in patients with sepsis have almost invariably been associated with a decrease in cardiac output and also, where measured, an increase in portal hypertension (*647, 648*). There has been no convincing evidence of improved outcome as yet, with the results of two larger trials in North America awaited. It has been suggested that selective iNOS inhibition may be beneficial as the protective effects of eNOS will not be altered. However, there is recent evidence suggesting downregulation of eNOS in sepsis (*649*). Combinations of NOS inhibitors and inhaled NO are being investigated to overcome the deleterious effect that NOS inhibitors have on pulmonary pressure. It may be that the ideal combination in treatment may include NOS inhibitors, inhaled NO plus selective NO donation both to maintain vulnerable microcirculations and to maintain antiplatelet effects. Clearly more work in this field is required.

11.8. NO and Cardiovascular Disease

The vasodepressor action of organic nitrates and nitrites has been known for over 100 years. BALARD synthesised amyl nitrite in 1844 and in 1867 BRUNTON described its use in the relief of angina pectoris (*650*). Glyceryl trinitrate (GTN) was synthesised shortly after amyl nitrite and was first tested in the treatment of angina in 1879 by MURRELL (*651*). Nitrovasodilators are now widely used in the treatment of angina, congestive cardiac failure and hypertensive emergencies. The mechanism of action of these drugs however was totally unknown until the 1970s when guanylate cyclase activation by GTN was demonstrated (*652*). The mediator of this activation is now known to be NO. NO formation from GTN, one of the most widely used nitrovasodilators, correlates with the GTN mediated vascular responses seen (*653, 654*). Continuous administration of GTN however causes tolerance to its effects within 24 hours, such that in clinical practice a "nitrate-free" period of at least 6–7 hours in every 24 hours is required to maintain its effectiveness. The mechanism of tolerance remains unclear but appears to be related to reduced biological activity of NO (*655*), rather than a reduced bioconversion of GTN to NO as has been previously proposed. GTN is predominantly venoselective, which is of benefit in the treatment of heart failure where there is increased preload, but this effect also causes postural hypotension, severe headache and has minimal anti-platelet effects. Sodium nitroprusside (SNP) is used especially for peri-operative control of blood pressure as it is short-acting with rapid onset, but it is a toxic agent producing increased plasma concentrations of cyanide. In view of the problems with GTN and SNP research has been directed at development of non-toxic NO-donors with no tolerance problem. The stability of these donors is also important. It has been suggested that the biological effects of nitrates and indeed endogenous NO may be mediated by the intermediate formation of nitrosothiols (*656*) and there is much interest in the clinical use of such compounds. This topic is described in section 6 of this review.

Over recent years the role of the endothelium in cardiovascular diseases has come to the fore. The role of NO as both an endothelium-derived vasodilator and an anti-thrombotic agent is increasingly recognised, with increasing implications in its role in the pathophysiology of disease states with endothelial dysfunction. The normal physiological function of NO appears to be protective in the cardiovascular system, though excessive production following the induction of iNOS may be harmful.

Atheroma may be considered to be a chronic inflammatory process resulting in the disease of vessel walls. NO may prevent atheroma formation, and indeed experimental evidence suggests that rats fed high dietary arginine show regression of fatty streaks and some reduced plaque formation. In experimental diabetes in rats there is good evidence that decreased vascular reactivity is due to abnormality in the NO pathway in large conduit arteries (*657*) and small resistance vessels (*658*).

The role of NO in hypertension remains contentious. Experimental work using NO inhibitors to investigate the role of NO in vascular tone has shown marked differences in differing vascular beds (*659*) and isolated vascular arteries (*660*) and has suggested that the increase in blood pressure following L-NNA administration is accounted for largely by increased peripheral resistance. Indirect evidence has suggested that the L-arginine-NO pathway may be impaired in established hypertension, but there are conflicting data from *in vitro* studies, with some groups noting impaired release of NO in hypertension (*661, 662*) whilst other groups have not confirmed this (*663*). Similarly there is conflicting evidence from *in vivo* studies, some showing decreased basal NO induced vascular tone in hypertensive rats (*664*), others showing no change (*658*), and still others showing enhanced NO release (*664, 665*). Development of atherosclerosis is increased in those with hypertension and all major adverse events associated with coronary heart disease (infarction, sudden death, angina, heart failure) are also increased in those with hypertension. Animal studies shown atherosclerotic plaques rarely develop solely as a result of hypertension, but are accelerated in those with hypertension and hypercholesterolemia (*666*). Clearly a fuller understanding of the precise role of NO in the endothelial dysfunction in these disease states will allow development of therapeutic modalities involving manipulation of the L-arginine-NO pathway.

11.9. NO and Interventional Cardiology

Balloon angioplasty is increasingly being used for the non-surgical treatment of coronary heart disease or as salvage technique in myocardial infarction in patients unsuitable for thrombolytic therapy. However in an elective situation, despite initial success rates of greater than 90%, restenosis (occlusion of the artery) occurs in 30–40% of cases within 6 weeks (*667*). In experimental models following balloon injury to vessel walls there is increased platelet aggregation and adhesion, with the proliferation and migration of smooth muscle cells into the intima where they continue to proliferate to heal the wound (*668*). Within a few days however, the aggregated platelets appear to disappear from the intimal surface, implying activation of anti-thrombotic mechanisms in the absence of regenerated endothelium (which regenerates over approximately 2 weeks). Recent evidence has shown impaired endothelial release of NO following experimental intimal damage which persisted despite histological regrowth of the endothelium (*668*). Since NO may inhibit vascular smooth muscle proliferation, the regrowth of this dysfunctional endothelium may contribute to an imbalance between growth-promoting and growth-inhibiting factors and promote restenosis.

Local NO delivered to the site of intimal damage in animals *in vivo* has been shown to improve the blood flow at the site (*669*). However, systemic administration of the compound did not confer any advantage. Locally administered single dose nitrosothiol appears also to prevent neointimal formation (*670*). *N*-Acetylcysteine, a drug which can be given orally and which can form a nitrosothiol, has been shown to reduce neointimal proliferation following balloon angioplasty (*671*). Thus nitrosothiols offer exciting potential in the prevention of restenosis following angioplasty and possible in the treatment of artherosclerosis *per se*. More results in this therapeutically exciting field are awaited, but there is clearly a potential role for NO-donors in therapeutic vascular intervention as well as other cardiovascular diseases such as angina, heart failure, hypertension, atherosclerosis and diabetes.

11.10. Pregnancy

Conrad and Vernier first showed increased plasma levels and urinary excretion of cGMP (*672*) and nitrate (*673*) in the pregnant rat suggesting altered NO production in pregnancy. This has subsequently been shown to be due to increased NOS since NOS inhibitor prevented this urinary excretion of nitrate and cGMP (*673*). The L-arginine-NO pathway appears to inhibit uterine contractility in the rat uterus (*674*) and also in human myometrium (*675*), with reduced NOS activity in late gestation (*676*) associated with a decrease in uterine relaxation responsiveness to NO at term which may be involved in the initiation of labour (*677*). NO is also important in regulating fetoplacental blood flow, which is a low pressure-high flow system (*678*). Infusion of an NOS inhibitor into the umbilical artery of pregnant sheep increases fetoplacental vascular resistance (*679*), and this has been backed up in human *in vitro* studies where L-NAME was shown to increases fetoplacental perfusion pressure (*680*).

11.11. Pre-Eclampsia

This condition is common during pregnancy and is heralded by the loss of unreponsiveness to Angiotensin II and subsequent development of vasoconstriction. It has been attributed to endothelial damage causing a deficiency of vasodilatory prostaglandins. However, it has been shown that prostaglandins do not mediate the attenuation of the systemic and renal vasopressor responsiveness seen in normal pregnancy. Blocking prostaglandin synthesis with indomethacin does not alter vascular resistance (*681*). There is now substantial evidence that endothelial cell dysfunction is associated with pre-eclampsia (*682*) but the causes of dysfunction remain unclear. One common symptom is excessive platelet activation and it has been suggested that this is caused by reduction in NO synthesis. This is substantiated by studies showing significantly lower plasma nitrite levels in pre-eclamptic patients, with a negative correlation between serum nitrite and diastolic blood pressure (*683*).

Sodium nitroprusside has been used to treat severe hypertension in pre-eclamptic patients (*684*), but this is an extremely potent vasodilator and can precipitate circulatory collapse in patients not adequately resuscitated with fluids to increase the depleted intravascular volume (*685*). There have been several studies in humans with pre-eclampsia using GTN, which is enzymatically metabolised into NO, for the management of hypertension but the results are confusing. GTN has little or no antiplatelet activity. Nitrosothiols such as GSNO, however, are both vasodilators and potent inhibitors of platelet activation and thus may be of more therapeutic benefit in pre-eclampsia. A recent case report using GSNO infusion in the treatment of HELLP syndrome, a rare variant of pre-eclampsia, has shown rapid improvement in the clinical conditions of the patient (*97*).

11.12. Nervous System

Basal release of NO induced by eNOS from cerebrovascular endothelium provides basal vasodilatory tone as demonstrated by evidence that NOS inhibitors reduce blood flow to the brain (*686*). NO also appears to be involved in the cerebrovascular vasodilatory response to hypercapnia, but evidence does not implicate its involvement in the vasodilation seen with hypotension or hypoxia. Neuronally released NO is implicated in the regional increase in blood flow associated with increased neural activity (687). It has been suggested that NO contributes to glutamate toxicity within the brain on the basis that NMDA receptor activation, which is seen in brain ischemia, activates nNOS to produce NO. Experimental data have shown a reduction in neurotoxicity mediated by NMDA-receptor activation following inhibition of NO synthesis *in vitro* (688). NO formation is increased in experimental models of stroke (689, 690), and several reports have demonstrated the beneficial effects of NOS inhibition (691–693), although this is not confirmed by all (694). It is likely that inhibition of iNOS is beneficial whereas inhibition of eNOS in endothelial cells leads to vasoconstriction and enhanced cerebral ischaemia. Indeed NO donors have been shown to be protective in various models of stroke (695, 696). It is likely that vasodilation and increased blood supply is the main protective factor but the anti-platelet properties of NO donors may also reduce vascular injury. Arterial NO-mediated vasodilation is impaired in hypertension, hypercholesterolemia, atherosclerosis and diabetes mellitus, all conditions associated with an increased risk of stroke. Patients with these etiological factors theoretically may benefit most from NO donors. These lines of evidence again highlight the need for both selective NOS inhibition to prevent unwanted side effects and directed NO donation. Some of the non-L-arginine-based NOS inhibitors which appear to demonstrate selective inhibition have been examined. 7-Nitroindazole, a putative iNOS inhibitor, has been used to reduce neuroinjury after occlusion and stroke (697, 698). Further work is awaited.

11.13. Bone

Increasing evidence implicates NO in the pathogenesis of bone diseases, principally those associated with cytokine activity. Both cNOS and iNOS are expressed by bone-derived cells. High concentrations of NO have been shown to have inhibitory effects on bone resorption and formation (699). Lower concentrations of NO however appear to be stimulatory to bone turnover, and SNAP enhances IL-1 induced bone resorption at low doses and suppresses it at high concentrations (699). There is strong evidence of increased NO production in inflammatory arthritis and, in animal models, this can be attenuated by NOS inhibitors (700, 701). Raised cytokines in inflammatory arthritis stimulate NO production from chondrocytes and synovial cells (*702, 703*) and studies in humans have shown elevated NO production in rheumatoid arthritis (*704, 705*). UEKI *et al.* (*706*) have investigated the relationship of NO to disease activity in rheumatoid arthritis. They showed that not only were serum levels of NO higher in parients with rheumatoid arthritis compared with controls or those with osteoarthritis, but also in those with active versus inactive disease. The serum levels were correlated to morning stiffness, number of swollen joints and C reactive protein (an inflammatory marker). This increase in serum NO may reflect increased NOS activity (*707*).

Whilst these findings may support the use of NOS inhibitors in the treatment of rheumatoid arthritis, the exact role of NO in joints remains uncertain, as it may have both pro-inflammatory and anti-inflammatory components to its activity. It has been suggested that NO is protective in acute experimentally-induced arthritis and deleterious in chronic inflammatory reactions but further investigation of NOS inhibitors as part of the therapeutic armoury of arthritis is warranted. IL-2 has been shown to attenuate the inflammatory response due to pro-inflammatory cytokines by inhibiting NO production and therefore represents therapeutic potential in all conditions where inflammation is prominent.

11.14. Conclusion

In view of the number of areas of human physiology where NO is implicated it is certain that NO-donors and NOS inhibitors will play an increasingly important role in clinical practice. NO gas is difficult to handle and readily contaminated and it may prove better to deliver NO to the pulmonary system with nebulised NO-donor drugs. In other situations, the challenge is to make NO-donor drugs tissue-selective as NO has some effect on almost every tissue it encounters. With NOS inhibitors the problem of selectivity between isoforms is also important. Future developments will be both interesting and exciting and, one hopes, of value to clinicians.

12. Addendum

Since this review was completed a number of important reports have emerged which throw further light on the structure and function of NOS. Firstly, in studies with the neuronal isozyme under single turnover conditions, STUEHR *et al.* have confirmed the NADPH stoichiometry for the Arg–NO pathway (*708*). In the case of nNOS, electron transfer from the reductase domain to the heme iron is completely dependent on the binding of CaM. Addition of a Ca^{2+} chelator (EDTA) results in dissociation of CaM from the enzyme, which has a lower affinity for this regulatory protein than iNOS or eNOS (*cf.* Section 10.9), thus preventing heme reduction. In this way STUEHR *et al.* were able to reduce resting ferric-heme nNOS to its ferrous state under anaerobic conditions with excess NADPH and then dissociate CaM to prevent further heme reduction. Addition of NHA and subsequent exposure to air led to the formation of one equivalent of Cit per mole of nNOS present. A single electron supplied by NADPH is, therefore, sufficient to catalyse one full round of NOS monooxygenation II, in a manner consistent with the stoichiometry shown in Fig. 5. Moreover, this stoichiometry rules out the formation of NO^- as the proximal product of NOS turnover, a possibility which has been the subject of recent debate (Section 8.1). The sufficiency of a single electron supplied by NADPH for NOS monooxygenation II also infers that this step in the Arg–NO pathway must be mediated by the complex between ferrous heme and O_2 [PPIX-Fe(III)-OO˙] as shown in Fig. 30, although the precise details of this mechanism still require confirmation. Interestingly, no products arose under these single turnover conditions in control experiments where Arg was substituted for NHA; this is consistent with the operation of oxoiron rather than peroxoiron chemistry in NOS monooxygenation I.

In related work from STUEHR'S group the formation and autoxidative decomposition of the NOS ferrous-dioxy complex [PPIX-Fe(II)$O_2\leftrightarrow$ PPIX-Fe(III)-OO˙] has been studied spectroscopically (*709*). These studies showed that the presence of bound H_4B significantly destabilises the complex, a feature that may be important for the catalytic role of the cofactor. As discussed in Section 10.6.3, there is a body of work which points to a role for H_4B which goes beyond that of an allosteric effector alone. New research from MAYER'S laboratories further strengthens this view (*710*). In this work the binding characteristics of radiolabelled H_4B and 4-aminotetrahydrobiopterin (4-aminoH_4B), a potent pterin-based inhibitor of NOS, were studied with the inducible isozyme. The dimeric structure of the enzyme was stabilised by 4-aminoH_4B in a manner apparently identical to its stabilisation by H_4B itself. Moreover, spectroscopic analysis suggested that the influence of 4-aminoH_4B on the heme coordination state is the same as that of H_4B. Unlike H_4B, however, the 4-amino derivative is resistant to enzymatic redox cycling, which hints at a redox role or some other highly specific function for the cofactor in NOS catalysis. The exact role of H_4B still remains unclear, however, and STUEHR *et al.* suggest (*708*) that it may not include a redox function in view of the close association between heme reduction and NO synthesis, the observed formation of a ferrous heme-dioxy complex and the lack of evidence for H_4B oxidation or redox cycling. New limited proteolysis studies with iNOS have also recently emerged and allowed identification of a 49-residue sequence at the N-terminus of the enzyme (residues 66–114) which is required for H_4B binding and dimerisation (*711*).

A number of recent papers have elaborated on the nature of the interactions between CaM and NOS (*cf.* Sections 10.3 and 10.4). MARLETTA *et al.* have studied the impact of the four Ca^{2+}-binding sites in CaM on electron transfer in nNOS using mutant CaM proteins that lack the ability to bind one or more Ca^{2+} ions (*712*). These studies demonstrated that binding of CaM to NOS is of itself not sufficient to promote electron transfer, but that interactions between the enzyme and specific residues in the Ca^{2+}-binding sites of CaM are essential. A model was proposed for the binding of CaM to NOS in order to account for the influence of particular Ca^{2+}-binding sites on electron transfer within the reductase domain and from the reductase domain to the oxygenase domain of NOS.

Three groups (*713–715*) have shown recently that membrane localisation of eNOS at the plasma membrane caveolae of endothelial cells involves direct interaction between the enzyme and caveolin-1, an important membrane protein component of the caveolae. The binding of caveolin-1 to eNOS, which inhibits the enzyme, is competitive and completely reversed by CaM binding. This suggests that caveolin-1 inhibits the enzyme by abrogating its activation by CaM and, therefore, that the enzyme may be subject to negative as well as positive allosteric regulation. Thus, VENEMA *et al.* (*713*) have suggested that the interaction between eNOS and caveolin-1 may provide a mechanism for deactivation of the enzyme after its stimulation by elevated intracellular Ca^{2+} levels. The same group have found the association of eNOS with caveolin-1 to be modulated by tyrosine phosphorylation of the latter (*716*).

The first NOS crystal structure, for a truncated murine iNOS oxygenase domain (residues 115–498; $iNOS_{ox}$ $\Delta114$) complexed with imidazole and aminoguanidine, has now been reported by TAINER, STUEHR and co-workers (*717*). The protein, which lacks the entire C-terminal reductase domain and the first 114 residues at the N-terminus, was crystallised as a heme-containing monomer without H_4B. The distal face of the heme is directed towards a large, exposed cavity which accommodates two imidazole ligands – one (IM1) coordinated directly to the heme iron centre and the other (IM2) lying above the plane of the heme over the edge of its A ring, Fig. 56. IM1 makes no hydrogen bonds to

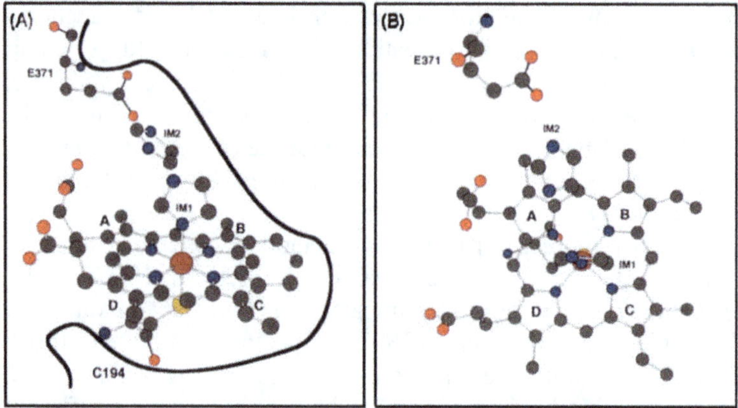

Fig. 56. Representation of the heme distal pocket of iNOSox D114 complexed with imidazole; (A) viewed from the C/D edge of the heme; (B) viewed from above. The distal pocket is defined by a canopy of mainly hydrophobic residues which opens to the exterior on the A/D edge of the heme. Two imidazole ligands (IM1 and IM2) are accommodated within the cavity. IM1 is bound by coordination to the heme iron; IM2 is located to the rear of IM1 above the porphyrin plane approximately over the edge of ring A. IM2 hydrogen bonds to E371, the glutamic acid residue which binds the Arg guanidine function in NOS

protein residues whereas IM2 is hydrogen bonded to E371, the important glutamic acid residue which has been shown to bind the Arg guanidine group at the catalytic centre (*298, 299*). The position of IM2 probably locates the Arg binding site – approximately over the heme A and B rings. Hydrophobic residues encircle the edge of the porphyrin ring and form an arc over the heme distal pocket (see Fig. 57); a peptide loop containing the proximal heme thiolate ligand (C194) closes the other face of the porphyrin. The heme is also sandwiched between tryptophan (W188) and phenylalanine (F363) residues which stack on the proximal and distal faces of the heme below and above the C ring respectively. In this way the proximal face of the heme is screened and an arced canopy that stretches over the heme distal pocket from its B and C rings at the rear. The cavity opens at the front above the heme A and D rings, with the propionate side chains of the latter extending out from the protein into solvent. The immediate vicinity of the porphyrin is dominated, therefore, by hydrophobic residues, with the exception of E371, and these residues are highly conserved across the three NOS isoforms.

The existence of two adjacent but distinct imidazole-binding sites was unexpected and perhaps has contributed to conflicting observations regarding the competitive or noncompetitive relationship between imidazole and substrate binding in the catalytic site (see Section 10.5.3). Thus,

```
HUMAN   nNOS                      HPSQHARRPEDVRTK·GQLFPLAKEFIDQYYSSIKRFGSKAHMERLEEVNKEIDTTSTY   0399 nNOS (human)
HUMAN   eNOS                      FPRKLQGRPSPGPPAPEQLLSQARDFINQYYSSIKRSGSQAHEQRLQEVEAEVAATGTY   0163 eNOS (human)
HUMAN   iNOS              0121    TPKSLTRGPRDKPTPPDELLPQAIEFVNQYYGSFKEAKIEEHLARVEAVTKEIETTGTY   0179 iNOS (human)
MURINE  iNOS              0115    NPKSLTRGPRDKPTPLEELLPHAIEFINQYYGSFKEAKIEEHLARLEAVTKEIETTGTY   0173 iNOS (murine)

                   1
QLKDTELIYGAKHAWRNASRCVGRIQWSKLQVFDARDCTTAHGMFNYICNHVKYATNKGNLRSAITIFPQRTDGKHDFRVWNSQLIRYAGYKQPDGSTLG   0499 nNOS (human)
QLRESELVFGAKQAWRNAPRCVGRIQWGKLQVFDARDCRSAQEMFTYICNHIKYATNRGNLRSAITVFPQRCPGRGDFRIWNSQLVRYAGYRQQDGSVRG   0263 eNOS (human)
QLTGDELIFATKQAWRNAPRCIGRIQWSNLQVFDARSCSTAREMFEHICRHVRYSTNNGNIRSAITVFPQRSDGKHDFRVWNAQLIRYAGYQMPDGSIRG   0279 iNOS (human)
QLTLDELIFATKMAWRNAPRCIGRIQWSNLQVFDARNCSTAQEMFQHICRHILYATNNGNIRSAITVFPQRSDGKHDFRLWNSQLIRYAGYQMPDGTIRG   0273 iNOS (murine)
                *    *****   *   *                          † †                ¶¶¶¶  ¶¶

                                                                                          2
DPANVQFTEICIQQGWKPPRGRFDVLPLLLQANGNDPELFQIPPELVLEVPIRHPKFEWFKDLGLKWYGLPAVSNMLLEIGGLEFSACPFSGWYMGTEIG   0599 nNOS (human)
DPANVEITELCIQHGWTPGNGRFDVLPLLLQAPDEPPELFLLPPELVLEVPLEHPTLEWFAALGLRWYALPAVSNMLLEIGGLEFPAAPFSGWYMSTEIG   0363 eNOS (human)
DPANVEFTQLCIDLGWKPKYGRFDVVPLVLQANGRDPELFEIPPDLVLEVAMEHPKYEWFRELELKWYALPAVANMLLEVGGLEFPGCPFNGWYMGTEIG   0379 iNOS (human)
DAATLEFTQLCIDLGWKPRYGRFDVLPLVLQADGQDPEVFEIPPDLVLEVTMEHPKYEWFQELGLKWYALPAVANMLIEVGGLEFPACPFNGWYMGTEIG   0373 iNOS (murine)
¶¶¶¶¶¶                                                      ¶¶  ††† ¶†              ††††††  †¶¶

VRDYCDNSRYNILEEVAKKMNLDMRKTSSLWKDQALVEINIAVLYSFQSDKVTIVDHHSATESFIKHMENEYRCRGGCPADWVWIVPPMSGSITPVFHQE   0699 nNOS (human)
TRNLCDPHRYNILEDVAVCMDLDTRTTSSLWKDKAAVEINVAVLHSYQLAKVTIVDHHAATASFMKHLENEQKARGGCPADWAWIVPPISGSLTPVFHQE   0463 eNOS (human)
VRDFCDVQRYNILEEVGRRMGLETHKLASLWKDQAVVEINIAVLHSFQKQNVTIMDHHSAAESFMKYMQNEYRSRGGCPADWIWLVPPMSGSITPVFHQE   0479 iNOS (human)
VRDFCDTQRYNILEEVGRRMGLETHTLASLWKDRAVTEINVAVLHSFQKQNVTIMDHHTASESFMKHMQNEYRARGGCPADWIWLVPPVSGSITPVFHQE   0473 iNOS (murine)
¶¶                                                          *

MLNYRLTPSFEYQPDPWNTHVWKGT                                                                             0724 nNOS (human)
MVNYFLSPAFRYQPDPWKGSAAKGT                                                                             0488 eNOS (human)
MLNYVLSPFYYYQVEAWKTHVWQDE                                                                             0504 iNOS (human)
MLNYVLSPFYYYQIEPWKTHIWQNE                                                                             0498 iNOS (murine)
        †  †¶¶  ¶  ¶
```

Fig. 57. Primary structure and sequence alignment for the three human NOS isoforms and murine iNOS$_{ox}$ Δ114. Marked residues are as follows: **1**, Cys residue that contributes proximal axial thiolate ligand for heme; **2**, glutamic acid residue that binds substrate's guanidine;*, residues bordering the heme on its edges and proximal face; †, distal pocket residues in the immediate region surrounding the heme; ¶, distal pocket residues in the extended region surrounding the heme

IM2 (but not IM1) might be expected to compete for the substrate-binding site. In the light of the $iNOS_{ox}$ $\Delta114$ crystal structure, Tainer *et al.* have suggested that dual-function inhibitors, which simultaneously bind to E371 and the heme iron, may possess enhanced binding affinity, blocking binding of both Arg and O_2.

Interestingly, the polypeptide topology of $iNOS_{ox}$ $\Delta114$ matches no other known protein, although clearly the high degree of sequence conservation across different NOS isoforms (Fig. 57) suggests that all three isozymes share a common structure. Of particular significance is the fact that the NOS oxygenase domain differs markedly in its tertiary structure, heme location and heme environment from known cytochrome P450s, even though both contain a thiolate-ligated heme. In the case of NOS, the thiolate forms two hydrogen bonds–to the protein backbone and to the indolic NH of a tryptophan residue (W188)–which are expected to lower the electron density on the sulfur atom. This feature of the NOS structure may have important ramifications for the mechanism of NOS catalysis. NOS monooxygenations I and II are generally thought to be mediated respectively by oxoiron and peroxoiron heme intermediates (see Sections 9.2 and 9.3). The propensity for cleavage of the O–O bond in the ferric heme peroxide complex [PPIX-Fe(III)-OOH] leading to formation of the oxoiron intermediate [PPIX$^{+\cdot}$-Fe(IV)=O] will be, therefore, an important controlling factor in the mechanism of NOS catalysis. In cytochrome P450 enzymes the strong donor thiolate ligand assists scission of the O–O bond. Thus, a reduction in the electron density of the axial thiolate in NOS, brought about by hydrogen bonding, may moderate the tendency for O–O bond cleavage and thereby permit peroxoiron mediation of NOS monooxygenation II.

As noted in Section 9.5, the delivery of protons to the catalytic centre may also affect the balance controlling the operation of peroxoiron and oxoiron chemistry in NOS catalysis. Tainer, Stuehr and co-workers noted a lack of structured water molecules in the heme distal pocket of the $iNOS_{ox}$ $\Delta114$ crystal structures which would also be consistent with an attenuated O–O bond cleaving activity and, therefore, perhaps allow a peroxoiron mediated step for NOS monooxygenation II. Finally, the presence of aromatic residues in the vicinity of the heme, most notably W188 and F363, may also influence the balance of peroxoiron versus oxoiron chemistry in NOS catalysis, for example, by charge transfer stabilisation of the porphyrin π-radical cation in the PPIX$^{+\cdot}$-Fe(IV)=O intermediate.

The mechanism-based NOS inhibitor aminoguanidine (AG) was found to displace IM2, binding above the edge of the heme A ring and A/B meso carbon, when soaked into the crystals of imidazole-complexed $iNOS_{ox}$ $\Delta114$. Binding of AG is facilitated by hydrogen bonds to E371, leaving one of the terminal guanidino nitrogen atoms directed towards the heme centre. In addition to lying in a position that would allow its oxidation TAINER *et al.* propose that substrate bound in the same position as AG could assist in activation of O_2 during NOS catalysis.

Superimposition of Arg on the NOS-bound AG suggested that the α-amino acid moiety of the substrate may lie at the edge of the heme distal pocket which is exposed to solvent in the monomeric $iNOS_{ox}$ $\Delta114$ structure.* The side of the protein on which the distal pocket opens was found to possess large areas of exposed hydrophobic residues which TAINER *et al.* suggest might contribute to the NOS dimer interface. This supposition is futher supported by the fact that known dimer-destabilising mutations also map to this face of the protein. The fact that the Arg amino acid moiety can tentatively be positioned near the dimer interface may explain why substrate/H_4B binding promotes dimerisation (see Section 10.7.). Dimerisation would be expected to bury the heme distal pocket and exclude bulk solvent. This idea is borne out by earlier analysis of the heme spin state and by ligand binding studies, discussed in Section 10.7, which suggested that the distal heme pocket of NOS subunits becomes exposed to solvent on dissociation of the dimeric protein (*550, 551*). The primary function of NOS dimerisation, therefore, may be to control the heme redox potential † allowing oxygen activation; this would explain why NOS is catalytically active only in the dimeric state. NOS dimerisation, then, contributes to the elaborate and elegant regulatory mechanisms which control NOS activity.

One important question that it is necessary to ask concerns the degree to which the structure of dimeric NOS, possessing bound H_4B and Arg, deviates from the crystal structure of $iNOS_{ox}$ $\Delta114$. As NOS is catalytically active only in its dimeric state, and dimerisation is promoted by the mutually cooperative binding of Arg and H_4B, it is reasonable to assume that Arg/H_4B binding and dimerisation induce significant conformational changes in the protein. The $iNOS_{ox}$ $\Delta114$ structure, however, is

---

* The conformation of Arg superimposed in these modelling studies was not described by TAINER *et al.* Recent work from the GLAXOWellcome Research Laboratories, however, suggests that the substrate (or at least Arg-analogue inhibitors of NOS) adopt a folded rather than an extended side chain conformation (*718*).

† The burying of a metalloporphyrin within a hydrophobic environment is known to have a profound influence on its redox potential. Thus, the more charged Fe(III) state of iron porphyrins sequestered in a dendritic core is destabilised relative to that of porphyrins for which there is free exposure to external bulk solvent; this can result in a strong shift of the redox potential to a more positive value (*719*). The heme prosthetic group of cytochromes P450 is buried in the core of the enzymes.

monomeric and devoid of H_4B. Interestingly, aryldiazene labelling studies carried out by ORTIZ de MONTELLANO *et al.* (*499, 500*), discussed in Section 10.5.3, suggest that binding of H_4B to NOS restricts the overall cavity size of the distal pocket and particularly so over the heme D ring. How closely, then, do the positions of IM2 and AG with respect to the heme centre in the $iNOS_{ox}$ $\Delta114$ crystal structures correspond to the location and orientation of the substrate guanidine function during NOS catalysis? Forthcoming crystal structures for full-length NOS_{ox} domains should solve this issue.

Acknowledgements

The authors wish to express their gratitude to the following people for helpful discussions during the preparation of this manuscript: Dr W. Alderton (GlaxoWellcome Research and Development, Stevenage, UK), Prof. T. L. Poulos (University of California, Irvine, USA), Prof. D. J. Stuehr (The Cleveland Clinic, Ohio, USA), Prof. A.-L. Tsai (University of Texas Medical School, Houston, USA), Prof. R. C. Venema (Medical College of Georgia, USA), Dr. K. Wong (Merck Research Laboratories, New Jersey, USA), Dr. R. J. Young (GlaxoWellcome Research and Development, Stevenage, UK). We thank also the early pioneers working on the roles of NO in mammalian physiology for providing so much intellectual excitement. We thank Fiona Campbell and Laura Hills for their contribution to the preparation of the manuscripts.

References

1. WINK, D.A., J.F. DARBYSHIRE, R.W. NIMS, J.E. SAAVEDRA, and P.C. FORD: Reactions of the Bioregulatory Agent Nitric Oxide in Oxygenated Aqueous Media – Determination of the Kinetics for Oxidation and Nitrosation by Intermediates Generated in the NO / O_2 Reaction. Chem. Res. Toxicol. **6**, 23 (1993).

2. AWAD, H.H., and D.M. STANBURY: Autoxidation of NO in Aqueous Solution. Int. J. Chem. Kinet. **25**, 375 (1993).

3. MURAD, F., W.P. ARNOLD, C.K. MITTAL, and J.M. BRAUGHLER: Properties and Regulation of Guanylate Cyclase and Some Proposed Functions for Cyclic GMP. Advances in Cyclic Nucleotide Research **11**, 175 (1979).

4. FURCHGOTT, R.F., and J.V. ZAWADZKI: The Obligatory Role of Endothelial Cells in the Relaxation of Arterial Smooth Muscle by Acetylcholine. Nature **288**, 373 (1980).

5. PALMER, R.M.J., A.G. FERRIGE, and S. MONCADA: Nitric Oxide Release Accounts for the Biological Activity of Endothelium-Derived Relaxing Factor. Nature **327**, 524 (1987).

6. IGNARRO, L.J., G.M. BUGA, K.S. WOOD, R.E.BYRNS, and G. CHAUDHURI: Endothelium-Derived Relaxing Factor Produced and Released from Artery and Vein is Nitric Oxide. Proc. Natl. Acad. Sci. USA **84**, 9265 (1988).

7. PALMER R.M.J., D.S. ASHTON, and S. MONCADA: Vascular Endothelial Cells Synthesize Nitric Oxide from L-Arginine. Nature **333**, 664 (1988).
8. SCHMIDT, H.H.H.W., H. NAU, W. WITTFOHT, J. GERLACH, K.E. PRESCHER, M.M. KLEIN, F. NIROOMAND, and E. BOHME: Arginine is a Physiological Precursor of Endothelium-Derived Nitric Oxide. Eur. J. Pharmacol. **154**, 213 (1988).
9. LEONE, A.M., R.M.J. PALMER, R.G. KNOWLES, P.L. FRANCIS, D.S. ASHTON, and S. MONCADA: Constitutive and Inducible Nitric Oxide Synthases Incorporate Molecular Oxygen into Both Nitric Oxide and Citrulline. J. Biol. Chem. **266**, 23790 (1991).
10. CRAVEN, P.A., and F.R. DeRUBERTIS: Requirement for Heme in the Activation of Purified Guanylate Cyclase by Nitric Oxide. Biochim. Biophys. Acta **745**, 310 (1983).
11. WALDMAN, S.A., and F. MURAD: Cyclic GMP Synthesis and Function. Pharmacol. Rev. **39**,163 (1987).
12. RADOMSKI, M.W., R.M.J. PALMER, and S. MONCADA: Comparative Pharmacology of Endothelium-Derived Relaxing Factor, Nitric Oxide and Prostacyclin in Platelets. Brit. J. Pharmacol. **92**, 181 (1987).
13. RADOMSKI, M.W., R.M.J. PALMER, and S. MONCADA: The Anti-Aggregating Properties of Vascular Endothelium – Interactions Between Prostacyclin and Nitric Oxide. Brit. J. Pharmacol. **92**, 639 (1987).
14. RADOMSKI, M.W., R.M.J. PALMER, and S. MONCADA: Characterization of the L-Arginine-Nitric Oxide Pathway in Human Platelets. Brit. J. Pharmacol. **101**, 325 (1990).
15. RADOMSKI, M.W., D.D. REES, A. DUTRA, and S. MONCADA: *S*-Nitroso-Glutathione Inhibits Platelet Activation *In Vitro* and *In Vivo*. Brit. J. Pharmacol. **107**, 745 (1992).
16. MENDELSOHN, M.E., S. O'NEILL, D. GEORGE, and J. LASCALZO: Inhibition of Fibrinogen to Human Platelets by S-Nitroso-N-acetylcysteine. J. Biol. Chem. **265**, 19028 (1990).
17. GORDGE, M.P., J.S. HOTHERSALL, G.S. NEILD, and A.A. NORONHA DUTRA: Role of a Copper(I)-Dependent Enzyme in the Anti-Platelet Action of S-Nitroso- glutathione. Brit. J. Pharamacol. **118**, 533 (1996).
18. FARIAS-EISNER, R., M.P. SHERMAN, E. AEBERHARD, and G. CHAUDHURI: Nitric Oxide is an Important Mediator for Tumoricidal Activity *In Vivo*. Pharmacology **91**, 9407 (1994).
19. WAGNER, D.A., V.R. YOUNG, and S.R. TANNENBAUM: Mammalian Nitrate Biosynthesis – Incorporation of (NH_3)-N-15 into Nitrate is Enhanced by Endotoxin Treatment. Proc. Natl. Acad. Sci. USA **80**, 4518 (1983).
20. HEGESH, E., and J. SHILOAH: Blood Nitrates and Infantile Methemoglobinemia. Clin. Chim. Acta **125**, 107 (1982).
21. STUEHR, D.J., and M.A. MARLETTA: Synthesis of Nitrite and Nitrate in Murine Macrophage Cell Lines. Cancer Res. **47**, 5590 (1987).
22. HIBBS, J.B., Z. VAVRIN, and R.R. TAINTOR: L-Arginine Is Required for Expression of the Activated Macrophage Effector Mechanism Causing Selective Metabolic Inhibition in Target Cells. J. Immunol. **138**, 550 (1987).
23. MARLETTA, M.A., P.S. YOON, R. IYENGAR, C.D. LEAF, and J.S. WISHNOK: Macrophage Oxidation of L-Arginine to Nitrite and Nitrate – Nitric Oxide Is an Intermediate. Biochemistry **27**, 8706 (1988).
24. HIBBS, J.B., R.R. TAINTOR, Z. VAVRIN, and E.M. RACHLIN: Nitric Oxide – A Cytotoxic Activated Macrophage Effector Molecule. Biochem. Biophys. Res. Commun. **157**, 87 (1988).
25. STUEHR, D.J., S.S. GROSS, I. SAKUMA, R. LEVI, and C.F. NATHAN: Activated Murine Macrophages Secrete a Metabolite of Arginine with the Bioactivity of Endothelium-Derived Relaxing Factor and the Chemical Reactivity of Nitric Oxide. J. Exp. Med. **169**, 1011 (1989).

26. GLAUSER, M.P., G. ZANETTI, J.-D. BAUMGARTNER, and J. COHEN; Septic Shock: Pathogenesis. Lancet **338**, 732 (1991).

27. PETROS, A., G. LAMB, A. LEONE, S. MONCADA, D. BENNETT, and P. VALLANCE: Effects of a Nitric-Oxide Synthase Inhibitor in Humans with Septic Shock. Cardiovascular Res. **28**, 34 (1994).

28. AFZAL, M., and J.C. WALTON: Oxidation of Retinyl Acetate and Analogues by Nitric Oxide and Nitrogen Dioxide. Bioorgan & Med. Chem. Letters **6**, 2329 (1996).

29. NGUYEN, T., D. BRUNSON, C.L. CRESPI, B.W. PENMAN, J.S. WISHNOK, and S.R. TANNENBAUM: DNA Damage and Mutation in Human Cells Exposed to Nitric Oxide *In Vitro*. Proc. Natl. Acad. Sci. USA **89**, 3030 (1992).

30. BECKMAN, J..S, T.W. BECKMAN, J. CHEN, P.A. MARSHALL, and B.A. FREEMAN: Apparent Hydroxyl Radical Production by Peroxynitrite – Implications for Endothelial Injury from Nitric Oxide and Superoxide. Proc. Natl. Acad. Sci. USA **87**, 1620 (1990).

31. GRYGLEWSKI, R.J., R.M.J. PALMER, and S. MONCADA: Superoxide Anion Is Involved in the Breakdown of Endothelium-Derived Vascular Relaxing Factor. Nature **320**, 454 (1986).

32. ISCHIROPOULOS, H., L. ZHU, J. CHEN, M. TSAI, J.C. MARTIN, C.D. SMITH, and J.S. BECKMAN: Peroxynitrite-Mediated Tyrosine Nitration Catalyzed by Superoxide Dismutase. Arch. Biochem. Biophys. **298**, 431 (1992).

33. ISCHIROPOULOS, H., L. ZHU, and J.S. BECKMAN: Peroxynitrite Formation from Macrophage-Derived Nitric Oxide. Arch. Biochem. Biophys. **298**, 446 (1992).

34. RADI, R., J.S. BECKMAN, K.M. BUSH, and B.A. FREEMAN: Peroxynitrite Oxidation of Sulfhydryls – The Cytotoxic Potential of Superoxide and Nitric Oxide. J. Biol. Chem. **266**, 4244 (1991).

35. FUKUTO, J.M., and L.J. IGNARRO: *In Vivo* Aspects of Nitric Oxide (NO) Chemistry: Does Peroxynitrite (-OONO) Play a Major Role in Cytotoxicity? Accounts Chem. Res. **30**, 149 (1997).

36. BUTLER, A.R., T.J. RUTHERFORD, D.M. SHORT, and J.H. RIDD: Tyrosine Nitration and Peroxonitrite (Peroxynitrite) Isomerisation: N-15 CIDNP NMR Studies. Chem. Comm. 669 (1997).

37. BUTLER, A.R., C. GLIDEWELL, A.R. HYDE, and J.C. WALTON: Formation of Paramagnetic Mononuclear Iron Nitrosyl Complexes from Diamagnetic Dinuclear and Tetranuclear Iron Sulfur Nitrosyls – Characterization by Electron Paramagnetic Resonance Spectroscopy and Study of Thiolate and Nitrosyl Ligand Exchange Reactions. Polyhedron **4**, 797 (1985).

38. HENRY, Y., M. LEPOIVRE, J.C. DRAPIER, C. DUCROCQ, J.L. BOUCHER, and A. GUISSANI: EPR Characterization of Molecular Targets for NO in Mammalian Cells and Organelles. FASEB J. **7**, 1124 (1993).

39. MULSCH, A., P.I. MORDVINTCEV, A.F. VANIN, and R. BUSSE: Formation and Release of Dinitrosyl Iron Complexes by Endothelial Cells. Biochem. Biophys. Res. Commun. **196**, 1303 (1993).

40. VANIN, A.F., P.I. MORDVINTCEV, S. HAUSCHILDT, and A. MULSCH: The Relationship Between L-Arginine Dependent Nitric Oxide Synthesis, Nitrite Release and Dinitrosyl Iron Complex Formation by Activated Macrophages. Biochim. Biophys. Acta **1177**, 37 (1993).

41. LANCASTER, J.R., G. WERNER-FELMAYER, and H. WACHTER: Coinduction of Nitric Oxide Synthesis and Intracellular Nonheme Iron Nitrosyl Complexes in Murine Cytokine Treated Fibroblasts. Free Radical Biol. Med. **16**, 869 (1994).

42. DRAPIER, J.-C., H. HIRLING, J. WIETZERBIN, P. KALDY, and L.C. KÜHN: Biosynthesis of Nitric Oxide Activates Iron Regulatory Factor in Macrophages. EMBO J. **12**, 3643 (1993).

43. MARAJ, S.R., S. KHAN, X.Y. CUI, R. CAMMACK, C.L. JOANNOU, and M.N. HUGHES: Interactions of Nitric Oxide and Redox Related Species with Biological Targets. Analyst **120**, 699 (1995).

44. CAMPBELL, J.M., F. MCCRAE, J. REGLINSKI, R. WILSON, W.E. SMITH, and R.D. STURROCK: The Interaction of Sodium Nitroprusside with Peripheral White Blood Cells *In Vitro*: A Rationale for Cyanide Release *In Vivo*. Biochim. Biophys. Acta **1156**, 327 (1993).

45. SUNG, S.-S., C. GLIDEWELL, A.R. BUTLER, and R. HOFFMANN: Bonding in Nitrosylated Iron-Sulfur Clusters. Inorganic Chem. **24**, 3856 (1985).

46. STAMLER, J.S., D.I. SIMON, J.A. OSBORNE, M.E. MULLINS, O. JARAKI, T. MICHEL, D.J. SINGEL, and J. LOSCALZO: *S*-Nitrosylation of Proteins with Nitric Oxide – Synthesis and Characterization of Biologically Active Compounds. Proc. Natl. Acad. Sci. USA **89**, 444 (1992).

47. HAUSLADEN, A., C.T. PRIVALLE, T. KENG, J. DE ANGELO, and J.S. STAMLER: Nitrosative Stress: Activation of the Transcription Factor OxyR. Cell **86**, 719 (1996).

48. DE GROOTE, M.A., D. GRANGER, Y. XU, G. CAMPBELL, R. PRINCE, and F.C. FANG: Genetic and Redox Determinants of Nitric Oxide Cytotoxicity in a Salmonella Typhimurium Model. Proc. Natl. Acad. Sci. USA **92**, 6399 (1995).

49. SOUTHAM, E., and J. GARTHWAITE: Comparative Effects of Some Nitric Oxide Donors on Cylic GMP Levels in Rat Cerebellar Slices. Neuroscience Letters **130**, 107 (1991).

50. DEGUCHI, T., and M. YOSHIOKA: L-Arginine Identified as an Endogenous Activator for Soluble Guanylate Cyclase from Neuroblastoma Cells. J. Biol. Chem. **257**, 147 (1982).

51. GARTHWAITE, J., S.L. CHARLES, and R. CHESS-WILLIAMS: Endothelium-Derived Relaxing Factor Released on Activation of NMDA Receptors Suggests Role as Intercellular Messenger in the Brain. Nature **336**, 385 (1988).

52. KNOWLES, R.G., M. PALACIOS, R.M.J. PALMER, and S. MONCADA: Formation of Nitric Oxide from L-Arginine in the Central Nervous System: A Transduction Mechanism for Stimulation of the Soluble Guanylate Cyclase. Proc. Natl. Acad. Sci. USA **86**, 5159 (1989).

53. BREDT, D.S., and S.H. SNYDER: Isolation of Nitric Oxide Synthetase, a Calmodulin Requiring Enzyme. Proc. Natl. Acad. Sci. USA **87**, 682 (1990).

54. IADECOLA, C.: Bright and Dark Sides of Nitric Oxide in Ischemic Brain Injury. Trends in Neurosciences **20**, 132 (1997).

55. GILLESPIE, J.S., X.R. LIU, and W. MARTIN: The Effects of L-Arginine and N^G-Monomethyl-L-arginine on the Response of the Rat Anococcygeus Muscle to NANC Nerve Stimulation. Brit. J. Pharmacol. **98**, 1080 (1989).

56. RAMAGOPAL, M.V., and H.J. LEIGHTON: Effects of N^G-Monomethyl-L-arginine on Field Stimulation Induced Decreases in Cytosolic Ca^{2+} Levels and Relaxation in the Rat Anococcygeus Muscle. Eur. J. Pharmacol. **174**, 297 (1989).

57. GIBSON, A., S. MIRZAZADEH, A.J. HOBBS, and P.K. MOORE: L-N^G- Monomethylarginine and L-N^G-Nitroarginine Inhibit Nonadrenergic, Noncholinergic Relaxation of the Mouse Anococcygeus Muscle. Brit. J. Pharmacol. **99**, 602 (1990).

58. BREDT, D.S., P.M. HWANG, and S.H. SNYDER: Localization of Nitric Oxide Synthase Indicating a Neural Role for Nitric Oxide. Nature **347**, 768 (1990).

59. MARTIN, J., and J.S. GILLESPIE. In: Novel Peripheral Transmitter (C. Bell, ed.). Pergamon Press, N. Y., 1990, p. 65.

60. IGNARRO, L.J., P.A. BUSH, G.M. BUGA, K.S. WOOD, J.M. FUKUTO, and J. RAJFER: Nitric Oxide and Cyclic-GMP Formation Upon Electrical Field Stimulation Cause Relaxation of Corpus Cavernosum Smooth Muscle. Biochem. Biophys. Res. Commun. **170**, 843 (1990).

61. Williams, D.L.H.: S-Nitrosation and the Reactions of S-Nitroso Compounds. Chem. Soc. Rev. **14**, 171 (1985).

62. Fontecave, M., and J.L. Pierre: The Basic Chemistry of Nitric Oxide and Its Possible Biological Reactions. Bull. Soc. Chim. France **131**, 620 (1994).

63. Field, L., R.V. Dilts, R. Ravichandran, P.G. Lenhert, and G.E. Carnahan: An Unusually Stable Thionitrite from N-Acetyl-D,L-penicillamine; X-Ray Crystal and Molecular Structure of 2-(Acetylamine)-2-carboxy-1,1-dimethylethyl Thionitrite: J. Chem. Soc. Chem. Comm., 249 (1978).

64. Bainbrigge, N., A.R. Butler, and C.H. Görbitz: The Thermal Stability of S-Nitrosothiols: Experimental Studies and Ab Initio Calculations on Model Compounds. J. Chem. Soc. Perkin Trans. 2, 351 (1997).

65. Hart, T.W.: Some Observations Concerning the S-Nitroso and S-Phenylsulfonyl Derivatives of L-Cysteine and Glutathione. Tetrahedron Letters **26**, 2013 (1985).

66. Rheinboldt, H., and F. Mott: Über die Thermische Dissoziation der Alkylthionitrite. J. Prakt. Chem. **133**, 328 (1932).

67. Barrett, J., L.J. Fitzgibbons, J. Glauser, R.H. Still, and P.N.W. Young: Photochemistry of the S-Nitroso Derivatives of Hexane-1-thiol and Hexane-1,6-dithiol. Nature **211**, 848 (1966).

68. Sexton, D.J., A. Muruganandam, D,J, McKenney, and B. Mutus: Visible Light Photochemical Release of Nitric Oxide from S-Nitrosoglutathione: Potential Photochemotherapeutic Applications. Photochem. Photobiol. **59**, 463 (1994).

69. Mcaninly, J., D.L.H. Williams, S.C. Askew, A.R. Butler, and C. Russell: Metal Ion Catalysis in Nitrosothiol (RSNO) Decomposition. J. Chem. Soc. Chem. Comm. **23**, 1758 (1993).

70. Dicks, A.P., H.R. Swift, D.L.H. Williams, A.R. Butler, H.H. Al-Sa'doni, and B.G. Cox: Identification of Cu^+ as the Effective Reagent in Nitric Oxide Formation from S-Nitrosothiols (RSNO). J. Chem. Soc. Perkin Trans. 2, 481 (1996).

71. Williams. D.L.H.: The Mechanism of Nitric Oxide Formation from S-Nitrosothiols (Thionitrites). J. Chem. Soc. Chem. Comm. 1085 (1996).

72. Dicks, A.P., P.H. Beloso, and D.L.H. Williams: Decomposition of S-Nitrosothiols: The Effects of Added Thiols. J. Chem. Soc. Perkin Trans. 2, 1429 (1997).

73. Myersc, P.R., R.L. Minor, R. Guerra, J.N. Bates, and D.G. Harrison: Vasorelaxant Properties of the Endothelium-Derived Relaxing Factor More Closely Resemble S-Nitrosocysteine than Nitric Oxide. Nature **345**, 161 (1990).

74. Feelisch, M., M. Te Poel, R. Zamora, A. Deussen, and S. Moncada: Understanding the Controversy over the Identity of EDRF. Nature **368**, 62 (1994).

75. Gustafsson, L.E., A.M. Leone, M.G. Persson, N.P. Wiklund, and S. Moncada: Endogenous Nitric Oxide is Present in the Exhaled Air of Rabbits, Guinea Pigs and Humans. Biochem. Biophys. Res. Comm. **181**, 852 (1991).

76. Kharitonov, V.G., A.R. Sundquist, and V.S. Sharma: Kinetics of Nitrosation of Thiols by Nitric Oxide in the Presence of Oxygen. J. Biol. Chem. **270**, 28158 (1995).

77. Goldstein, S., and G. Czapski: Mechanism of the Nitrosation of Thiols and Amines by Oxygenated NO Solutions – The Nature of the Nitrosating Intermediates. J. Amer. Chem. Soc. **118**, 3419 (1996).

78. Jia, L., C. Bonaventura, J. Bonaventura, and J.S. Stamler: S- Nitrosohemoglobin – A Dynamic Activity of Blood Involved in Vascular Control. Nature **380**, 221 (1996).

79. Lancaster, J.R.: Simulation of the Diffusion and Reaction of Endogenously Produced Nitric Oxide. Proc. Natl. Acad. Sci. USA **91**, 8137 (1994).

80. Guo, F.H., H.R. De Raeve, H.R. Rice, D.J. Stuehr, F.B.M.J. Thunissen, and S.C. Erzerum: Continuous Nitric Oxide Synthesis by Inducible Nitric Oxide Synthase in

Normal Human Airway Epithelium *In Vivo*. Proc. Natl. Acad. Sci. USA **92**, 7809 (1995).

81. VALLANCE, P., J. COLLIER, and S. MONCADA: Effects of Endothelium-Derived Nitric Oxide on Peripheral Arteriolar Tone in Man. Lancet **2**, 997 (1989).

82. FURCHGOTT, R.F.: Annual Rev. Pharmacol. Toxicol. **24**, 175 (1984).

83. RUBANYI, G.M., J.C. ROMERO, and P.M. VAN HOUTTE: Flow-Induced Release of Endothelium-Derived Relaxing Factor. Am. J. Physiol. **250**, 1145 (1986).

84. STAMLER, J.S.: *S*-Nitrosothiols and the Bioregulatory Actions of Nitrogen Oxides Through Reactions with Thiol Groups. Current Topics in Microbiology and Immunology **196**, 19 (1995).

85. MELLION, B.T., L.J. IGNARRO, C.B. MYERS, E.H. OHLSTEIN, B.A. BALLOT, A.L. HYMAN, and P.J. KADOWITZ: Inhibition of Human Platelet Aggregation by *S*- Nitrosothiols. Heme Dependent Activation of Soluble Guanylate Cyclase and Stimulation of Cyclic GMP Accumulation. Mol. Pharmacol. **23**, 653 (1983).

86. MOYNIHAN, H.A., and S.M. ROBERTS: Preparation of Some Novel *S*-Nitroso Compounds as Potential Slow Release Agents of Nitric Oxide *In Vivo*. J. Chem. Soc. Perkin Trans. 1, 797 (1994).

87. MACALLISTER, R.J., A.L. CALVER, J. RIEZEBOS, J. COLLIER, and P. VALLANCE: Relative Potency and Arteriovenous Selectivity of Nitrovasodilators on Human Blood Vessels – An Insight into the Targeting of Nitric Oxide Delivery. J. Pharmacol. Exptl. Ther. **273**, 154 (1995).

88. MATHEWS, W.R., and S.W. KERR: Biological Activity of *S*-Nitrosothiols – The Role of Nitric Oxide. J. Pharmacol. Exptl. Ther. **267**, 1529 (1993).

89. GASTON, B., J.M. DRAZEN, A. JANSEN, D.A. SUGARBAKER, J. LOSCALZO, W. RICHARDS, and J.S. STAMLER: Relaxation of Human Bronchial Smooth Muscle by *S*-Nitrosothiols *In Vitro*. J. Pharmacol. Exptl. Ther. **268**, 978 (1994).

90. KOWALUK, E.A., and H.L. FUNG: Spontaneous Liberation of Nitric Oxide Cannot Account for *In Vitro* Vascular Relaxation by *S*-Nitrosothiols. J. Pharmacol. Exptl. Ther. **255**, 1256 (1990).

91. BAUER, J.A., and H.L. FUNG: Differential Hemodynamic Effects And Tolerance Properties of Nitroglycerin and an *S*-Nitrosothiol in Experimental Heart Failure. J. Pharmacol. Exptl. Ther. **256**, 249 (1991).

92. HOROWITZ, J.D., E.M. ANTMAN, B.H. LORELL, W.H. BARRY, and T.W. SMITH: Potentiation of the Cardiovascular Effects of Nitroglycerin by *N*-Acetylcysteine. Circulation **68**, 1247 (1983).

93. IGNARRO, L.J., H. LIPPTON, J.C. EDWARDS, W.H. BARICOS, A.L. HYMAN, P.J. KADOWITZ, and C.A. GRUETTER: Mechanism of Vascular Smooth Muscle Relaxation by Organic Nitrates, Nitrites, Nitroprusside and Nitric Oxide – Evidence for the Involvement of *S*-Nitrosothiols as Active Intermediates. J. Pharmacol. Exptl. Ther. **218**, 739 (1981).

94. SALAS, E., M.A. MORO, S. ASKEW, H.F. HODSON, A.R. BUTLER, M.W. RADOMSKI, and S. MONCADA: Comparative Pharmacology of Analogues of *S*- Nitroso-*N*-acetyl-D,L-penicillamine on Human Platelets. Brit. J. Pharmacol. **112**, 1071 (1994).

95. ASKEW, S.C., A.R. BUTLER, F.W. FLITNEY, G.D. KEMP, and I.L. MEGSON: Chemical Mechanisms Underlying the Vasodilator and Platelet Anti-Aggregating Properties of *S*-Nitroso-*N*-acetyl-D,L-penicillamine and *S*-Nitrosoglutathione. Biorg. & Med. Chem. **3**, 1 (1995).

96. LANGFORD, E.J., A.S. BROWN, R.J. WAINWRIGHT, A.J. DE BELDER, M.R. THOMAS, R.E.A. SMITH, M.W. RADOMSKI, J.F. MARTIN, and S. MONCADA: Inhibition of Platelet Activity by *S*-Nitrosoglutathione During Coronary Angioplasty. Lancet **344**, 1458 (1994).

97. DE BELDER, A., C. LEES, J. MARTIN, S. MONCADA, and S. CAMPBELL: Treatment of HELLP Syndrome with Nitric Oxide Donor. Lancet **345**, 124 (1995).

98. DICKS, A.P., and D.L.H. WILLIAMS: Generation of Nitric Oxide from S-Nitrosothiols Using Protein-Bound Cu^{2+} Sources. Chem. & Biol. **3**, 655 (1996).

99. DE MAN, J.G., B.Y. DE WINTER, G.E. BOECKXSTAENS, A.G. HERMAN, and P.A. PELCKMANS: Effect of Cu^{2+} on Relaxations to the Nitrergic Neurotransmitter, NO and S-Nitrosothiols in the Rat Gastric Fundus. Brit. J. Pharmacol. **119**, 990 (1996).

100. GORDGE, M.P., D.J. MEYER, J. HOTHERSALL, G.H. NEILD, N.N. PAYNE, and A. NORONHADUTRA: Copper Chelation-Induced Reduction of the Biological Activity of S-Nitrosothiols. Brit. J. Pharmacol. **114**, 1083 (1995).

101. AL-SA'DONI, H.H., I.L. MEGSONS, S. BISLAND, A.R. BUTLER, and F.W. FLITNEY: Neocuproine, a Selective Cu(I) Chelator, and the Relaxation of Rat Vascular Smooth Muscle by S-Nitrosothiols. Brit. J. Pharmacol. **121**, 1047 (1997).

102. GORREN, A.C.F., A. SCHRAMMEL, K. SCHMIDT, and B. MAYER: Decomposition of S-Nitrosoglutathione in the Presence of Copper Ions and Glutathione. Arch. Biochem. Biophys. **330**, 219 (1996).

103. KOMIYAMA, T., and K. FUJIMORI: Kinetic Studies of the Reaction of S-Nitroso-L-cysteine with L-Cysteine. Bioorg. Med. Chem. Letters **7**, 175 (1997).

104. MEGSON, I.L., I.R. GREIG, A.R. BUTLER, G.A. GRAY, and D.J. WEBB: Therapeutic Potential of S-Nitrosothiols as Nitric Oxide Donor Drugs. Scottish Med. J. **42**, 88 (1997).

105. RAMIREZ, J., L. YU, J. LI, P.G. BRAUNSCHWEIGER, and P.G. WANG: Glyco-S-nitrosothiols, a Novel Class of NO Donor Drugs. Bioorg. Med. Chem. Letters **6**, 2575 (1996).

106. FLITNEY, F.W., I.L. MEGSON, D.E. FLITNEY, and A.R. BUTLER: Iron-Sulphur Cluster Nitrosyls, a Novel Class of Nitric Oxide Generator: Mechanism of Vasodilator Action on Rat Isolated Tail Artery. Brit. J. Pharmacol. **107**, 842 (1992).

107. DEUSSEN, A., M. SONNTAG, and R. VOGEL: L-Arginine Derived Nitric Oxide – A Major Determinant of Uveal Blood Flow. Exp. Eye Res. **57**, 129 (1993).

108. NATHANSON, J.A.: Nitrovasodilators as a New Class of Ocular Hypotensive Agents. J. Pharmacol. Exp. Ther. **260**, 956 (1992).

109. SCHUMAN, J.S., K. ERICKSON, and J.A. NATHANSON: Nitrovasodilator Effects on Intraocular Pressure and Outflow Facility in Monkeys. Exp. Eye Res. **58**, 99 (1994).

110. NATHANSON, J.A., and M. MCKEE: Identification of an Extensive System of Nitric Oxide Producing Cells in the Ciliary Muscle and Outflow Pathway of the Human Eye. Invest. Ophthalmol. Vis. Sci. **36**, 1765 (1995).

111. WIEDERHOLT, M., A. STURM, and A. LEPPLEWIENHUES: Relaxation of Trabecular Meshwork and Ciliary Muscle by Release of Nitric Oxide. Invest. Ophthalmol. Vis. Sci. **35**, 2515 (1994).

112. HAUFSCHILD, T., E. NAVA, P. MEYER, J. FLAMMER, T.F. LUSCHER, and I.O. HAEFLIGER: Spontaneous Calcium Independent Nitric Oxide Synthase Activity in Porcine Ciliary Processes. Biochem. Biophys. Res. Com. **222**, 786 (1996).

113. GOUREAU, O., D. HICKS, and Y. COURTOIS: Human Retinal Pigmented Epithelial Cells Produce Nitric Oxide in Response to Cytokines. Biochem. Biophys. Res. Com. **198**, 120 (1994).

114. TAMM, E.R., C. FLUGELKOCH, B. MAYER, and E. LUTJENDRECOLL: Nerve Cells in the Human Ciliary Muscle – Ultrastructural and Immunocytochemical Characterization. Invest. Ophthalmol. Vis. Sci. **36**, 414 (1995).

115. TILTON, R.G., K. CHANG, J.A. CORBETT, T.P. MISKO, M.G. CURRIE, N.S. BORA, H.J. KAPLAN, and J.R. WILLIAMSON: Endotoxin-Induced Uveitis in the Rat is Attenuated by Inhibition of Nitric Oxide Production. Invest. Ophthalmol. Vis. Sci. **35**, 3278 (1994).

116. SCHMIDT, K.G., O. GEYER, and T.W. MITTAG: Choroidal Vascular Tone – Potential Biochemical Pharmacological Mechanisms. Invest. Ophthalmol. Vis. Sci. **34**, 826 (1993).

117. GOUREAU, O., M. LEPOIVRE, F. MASCARELLI, and Y. COURTOIS: Nitric Oxide Synthase Activity in Bovine Retina. Struct. Funct. Prot. **221**, 395 (1992).

118. SCHMIDT, H.H.H.W., G.D. GAGNE, M. NAKANE, J.S. POLLOCK, M.F. MILLER, and F. MURAD: Mapping of Neural Nitric Oxide Synthase in the Rat Suggests Frequent Colocalization with NADPH Diaphorase but not with Soluble Guanylyl Cyclase, and Novel Paraneural Functions for Nitrinergic Signal Transduction. J. Histochem. Cytochem. *40*, 1439 (1992).

119. GRIFFITH, O.W., and D.J. STUEHR: Nitric Oxides Synthases: Properties and Catalytic Mechanism. Annu. Rev. Physiol. **57**, 707 (1995).

120. FUKUTO, J.M., G.C. WALLACE, R. HSZIEH, and G. CHAUDHURI: Chemical Oxidation of N-Hydroxyguanidine Compounds: Release of Nitric Oxide, Nitroxyl and Possible Relationship to the Mechanism of Biological Nitric Oxide Generation. Biochem. Pharmac. **43**, 607 (1992).

121. FUKUTO, J.M., D.J. STUEHR, P.L. FELDMAN, M.P. BOVA, and P. WONG: Peracid Oxidation of an N-Hydroxyguanidine Compound: A Chemical Model for the Oxidation of N^{ω}-Hydroxy-L-arginine by Nitric Oxide Synthase. J. Med. Chem. **36**, 2666 (1993).

122. FUKUTO, J.M., K. CHIANG, R. HSZIEH, P. WONG, and G. CHAUDHAURI: The Pharmacological Activity of Nitroxyl: A Potent Vasodilator with Activity Similar to Nitric Oxide and or Endothelium Derived Relaxing Factor. J. Pharmacol. Exp. Ther. **263**, 546 (1992).

123. GRAETZEL, M., S. TANAGUCHI, and A. HENGLEIN: Pulse Radiolytic Study of Short-lived By-products of Nitric Oxide Reduction in Aqueous Systems. Ber. Bunsen-Ges. Phys. Chem. **74**, 1003 (1970).

124. SCHMIDT, H.H.H.W., H. HOFMANN, U. SCHINDLER, Z.S. SHUTENKO, D.D. CUNNINGHAM, and M. FEELISCH: No NO from NO Synthase. Proc. Natl. Acad. Sci. USA **93**, 14492 (1996).

125. ZAMORA, R., A. GRZESIOK, H. WEBER, and M. FEELISCH: Oxidative Release of Nitric Oxide Accounts for Guanylyl Cyclase Stimulating, Vasodilator and Antiplatetlet Activity of Piloty's Acid: A Comparison with Angeli's Salt. Biochem. J. **312**, 333 (1995).

126. FUKUTO, J.M., A.J. HOBBS, and L.J. IGNARRO: Conversion of Nitroxyl (HNO) to Nitric Oxide (NO) in Biological Systems: The Role of Physiological Oxidants and Relevance to the Biological Activity of HNO. Biochem. Biophys. Res. Commun. **196**, 707 (1993).

127. MURPHY M.E., and H. SIES: Reversible Conversion of Nitroxyl Anion to Nitric Oxide by Superoxide Dismutase. Proc. Natl. Acad. Sci. USA **88**, 10860 (1991).

128. DIERKS, E.A., and J.N. BURSTYN: Nitric Oxide (NO·), the only Nitrogen Monoxide Redox Form Capable of Activating Soluble Guanylyl Cyclase. Biochem. Pharmac. **51**, 1593 (1996).

129. MANSUY, D., and P. BATTIONI: Cytochrome P450 Model Systems. In: Metalloporphyrins in Catalytic Oxidations, Chapter 4 (1994).

130. PORTER, T.D., and M.N. COON: Cytochrome P-450: Multiplicity of Isoforms, Substrates, and Catalytic and Regulatory Mechanisms. J. Biol. Chem. **266**, 13469 (1991).

131. PORTER, T.D., T.W. BECK, and C.B. KASPER: NADPH-Cytochrome P-450 Oxidoreductase Gene Organization Correlates with Structural Domains of the Protein. Biochemistry **29**, 9814 (1990).

132. NARHI, L.O., and A.J. FULCO: Characterization of a Catalytically Self-Sufficient 119,000-Dalton Cytochrome P-450 Monooxygenase Induced by Barbiturates in *Bacillus megaterum*. J. Biol. Chem. **261**, 7160 (1986).

133. NARHI, L.O., and A.J. FULCO: Identification and Characterization of 2 Functional Domains in Cytochrome P-450$_{BM-3}$, a Catalytically Self-Sufficient Monooxygenase Induced by Barbiturates in *Bacillus megaterium*. J. Biol. Chem. **262**, 6683 (1987).

134. FULCO, A.J.: P450BM-3 and Other Inducible Bacterial P450 Cytochromes – Biochemistry and Regulation. Annu. Rev. Pharmac. Toxiol. **31**, 177 (1991).

135. RUETTINGER, R.T., L.P. WEN, and A.J. FULCO: Coding Nucleotide, 5' Regulatory, and Deduced Amino Acid Sequences of P450BM-3, a Single Peptide Cytochrome P-450 / NADPH P-450 Reductase from *Bacillus megaterium*. J. Biol. Chem. **264**, 10987 (1989).

136. KLATT, P., K. SCHMIDT, and B. MAYER: Brain Nitric Oxide Synthase Is a Hemoprotein. Biochem. J. **288**, 15 (1992).

137. MCMILLAN, K., D.S. BREDT, D.J. HIRSCH, S.H. SNYDER, J.E. CLARK, and B.S.S. MASTERS: Cloned, Expressed Rat Cerebellar Nitric Oxide Synthase Contains Stoichiometric Amounts of Heme, which Binds Carbon Monoxide, Proc. Natl. Acad. Sci. USA **89**, 11141 (1992).

138. MASTERS, B.S.S., K. MCMILLAN, E.A. SHETA, J.S. NISHIMURA, L.J. ROMAN, and P. MARTASEK: Neuronal Nitric Oxide Synthase, a Modular Enzyme Formed by Convergent Evolution: Structure Studies of a Cysteine Thiolate-Liganded Heme Protein that Hydroxylates L-Arginine to Produce NO as a Cellular Signal **10**, 552 (1996).

139. WHITE, K.A., and M.A. MARLETTA: Nitric Oxide Synthase Is a Cytochrome P-450 Type Hemoprotein. Biochemistry **31**, 6627 (1992).

140. BREDT, D.S., P.M. HWANG, C.E. GLATT, C. LOWENSTEIN, R.R. REED, and S.H. SNYDER: Cloned and Expressed Nitric Oxide Synthase Structurally Resembles Cytochrome P-450 Reductase. Nature **351**, 714 (1991).

141. BREDT, D.S., and S.H. SNYDER: Nitric Oxide: A Physiological Messenger Molecule. Annu. Rev. Biochem. **63**, 175 (1994).

142. XIE, Q.W., H.J. CHO, J. CALAYCAY, R.A. MUMFORD, K.M. SWIDEREK, T.D. LEE, A.H. DING, T. TROSO, and C. NATHAN: Cloning and Characterization of Inducible Nitric Oxide Synthase from Mouse Macrophages. Science **256**, 225 (1992).

143. SESSA, W.C.: The Nitric Oxide Synthase Family of Proteins. J. Vasc. Res. **31**, 131 (1994).

144. MAYER, B., M. JOHN, B. HEINZEL, E.R. WERNER, H. WACHTER, G. SCHULTZ, and E. BOHME: Brain Oxide Synthase Is a Biopterin- and Flavin-Containing Multifunctional Oxidoreductase. FEBS Letters **288**, 187 (1991).

145. STUEHR, D.J., H.J. CHO, N.S. KWON, M.F. WEISE, and C.F. NATHAN: Purification and Characterization of the Cytokine-Induced Macrophage Nitric Oxide Synthase: a FAD-Containing and FMN-Containing Flavoprotein. Proc. Natl. Acad. Sci. USA **88**, 7773 (1991).

146. HEVEL, J.M., K.A. WHITE, and M.A. MARLETTA: Purification of the Inducible Murine Macrophage Nitric Oxide Synthase. Identification as a Flavoprotein. J. Biol. Chem. **266**, 22789 (1991).

147. TAYEH, M.A., and M.A. MARLETTA: Macrophage Oxidation of L-Arginine to Nitric Oxide, Nitrite, and Nitrate–Tetrahydrobiopterin Is Required as a Cofactor. J. Biol. Chem. **264**, 19654 (1989).

148. KWON, N.S., C.F. NATHAN, and D.J. STUEHR: Reduced Biopterin as a Cofactor in te Generation of Nitrogen Oxides by Murine Macrophages J. Biol. Chem. **264**, 20496 (1989).

149. MAYER B., M. JOHN, and E. BOHME: Purification of a Ca^{2+} / Calmodulin-Dependent Nitric Oxide Synthase from Porcine Cerebellum: Cofactor Role of Tetrahydrobiopterin. FEBS Letters **277**, 215 (1990).

150. SCHMIDT, K., E.R. WERNER, B. MAYER, H. WACHTER, and W.R. KUKOVETZ: Tetrahydrobiopterin-Dependent Formation of Endothelium Derived Relaxing Factor (Nitric Oxide) in Aortic Endothelial Cells. Biochem. J. **281**, 297 (1992).

151. SCHMIDT, H.H.H.W., R.M. SMITH, M. NAKANE, and F. MURAD: Ca^{2+} / Calmodulin-Dependent NO Synthase Type I: a Biopteroflavoprotein with Ca^{2+} / Calmodulin-Independent Diaphorase and Reductase Activities. Biochemistry **31**, 3243 (1992).

152. HEVEL, J.M. and M.A. MARLETTA: Macrophage Nitric Oxide Synthase: Relationship Between Enzyme-Bound Tetrahydrobiopterin and Synthase Activity. Biochemistry **31**, 7160 (1992).

153. KLATT, P., B. HEINZEL, B. MAYER, E. AMBACH, G. WERNER-FLEMAYER, H. WACHTER, and E.R. WERNER: Stimulation of Human Nitric Oxide Synthase by Tetrahydrobiopterin and Selective Binding of the Cofactor. FEBS Letters **305**, 160 (1992).

154. SAKAI, N., S. KAUFMAN, and S. MILSTIEN: Tetrahydrobiopterin Is Required For Cytokine-Induced Nitric Oxide Production in a Murine Macrophage Cell Line (Raw-264). Molec. Pharmac. **43**, 6 (1993).

155. JANSSENS, S.P., A. SHIMOUCHI, T. QUERTERMOUS, D.B. BLOCH, and K.D. BLOCH: Cloning and Expression of a cDNA Encoding Human Endothelium Derived Relaxing Factor Nitric Oxide Synthase. J. Biol. Chem. **267**, 14519 (1992).

156. LAMAS, S., P.A. MARSDEN, G.K. LI, P. TEMPST, and T. MICHEL: Endothelial Nitric Oxide Synthase: Molecular Cloning and Characterization of a Distinct Constitutive Enzyme Isoform. Proc. Natl. Acad. Sci. USA **89**, 6348 (1992).

157. NISHIDA, K., D.G. HARRISON, J.A. NAVAS, A.A. FISHER, S.P. DOCKERY, M. UEMATSU, R.M. NEREM, R.W. ALEXANDER, and T.J. MURPHY: Molecular Cloning and Characterization of the Constitutive Bovine Aortic Endothelial Cell Nitric Oxide Synthase. J. Clin. Invest. **90**, 2092 (1992).

158. SESSA, W.C., J.K. HARRISON, C.M. BARBER, D. ZENG, M.E. DURIEUX, D.D. DANGELO, K.R. LYNCH, and M.J. PEACH: Molecular Cloning and Expression of a cDNA Encoding Endothelial Nitric Oxide Synthase, J. Biol. Chem. **267**, 15274 (1992).

159. LOWENSTEIN, C.J., C.S. GLATT, D.S. BREDT, and S.H. SNYDER: Cloned and Expressed Macrophage Nitric Oxide Synthase Contrasts with the Brain Enzyme. Proc. Natl. Acad. Sci. USA **89**, 6711 (1992).

160. LYONS, C.R., G.J. ORLOFF, and J.M. CUNNINGHAM: Molecular Cloning and Functional Expression of an Inducible Nitric Oxide Synthase from a Murine Macrophage Cell Line. J. Biol. Chem. **267**, 6370 (1992).

161. NAKANE, M., H.H.H.W. SCHMIDT, J.S. POLLOCK, U. FÖRSTERMANN, and F. MURAD: Cloned Human Brain Nitric Oxide Synthase Is Highly Expressed in Skeletal Muscle. FEBS Letters **316**, 175 (1993).

162. CHARLES, I.G., R.M.J. PALMER, M.S. HICKERY, M.T. BAYLISS, A.P. CHUBB, V.S. HALL, D.W. MOSS, and S. MONCADA: Cloning, Characterization, and Expression of a cDNA Encoding an Inducible Nitric Oxide Synthase from the Human Chondrocyte. Proc. Natl. Acad. Sci. USA **90**, 11419 (1993).

163. GELLER, D.A., C.J. LOWENSTEIN, R.A. SHAPIRO, A.K. NUSSLER, M. DISIL VIO, S.C. WANG, D.K. NAKAYAMA, R.L. SIMMONS, S.H. SNYDER, and T.R. BILLIAR: Molecular Cloning and Expression of Inducible Nitric Oxide Synthase from Human Hepatocytes. Proc. Natl. Acad. Sci. USA **90**, 3491 (1993).

164. SHERMAN, P.A., V.E. LAUBACH, B.R. REEP, and E.R. WOOD: Purification and cDNA Sequence of an Inducible Nitric Oxide Synthase from a Human Tumor Cell Line. Biochemistry **32**, 11600 (1993).

165. NUNOKAWA, Y., N. ISHIDA, and S. TANAKA: Cloning of Inducible Nitric Oxide Synthase in Rat Vascular Smooth Muscle Cells. Biochem. Biophys. Res. Commun. **191**, 89 (1993).

166. STUEHR, D.J., and M. IKEDA-SAITO: Spectral Characterization of Brain and Macrophage Nitric Oxide Synthases: Cytochrome P-450-Like Hemeproteins that Contain a Flavin Semiquinone Radical. J. Biol. Chem. **267**, 20547 (1992).

167. MARLETTA, M.A.: Nitric Oxide Synthase Structure and Mechanism. J. Biol. Chem. **268**, 12231 (1993).

168. KNOWLES, R.G., and S. MONCADA: Nitric Oxide Synthases in Mammals. Biochem. J. **298**, 249 (1994).

169. MASTERS, B.S.S.: Nitric Oxide Synthases: Why So Complex? Annu. Rev. Nutrition **14**, 131 (1994).

170. MAYER, B.: Molecular Characteristics and Enzymology of Nitric Oxide Synthase and Soluble Guanylyl Cyclase in the CNS. Semin. Neurosci. **5**, 197 (1993).

171. LIU, Q., and S.S. GROSS: Binding Sites of Nitric Oxide Synthases. Methods Enzymol. **268**, 311 (1996).

172. BAEK, K.J., B.A. THIEL, S. LUCAS, and D.J. STUEHR: Macrophage Nitric Oxide Synthase Subunits: Purification, Characterization, and Role of Prosthetic Groups and Substrate in Regulating Their Association into a Dimeric Enzyme. J. Biol. Chem. **268**, 21120 (1993).

173. MAYER, B., and E.R. WERNER: In Search of a Function for Tetrahydrobiopterin in the Biosynthesis of Nitric Oxide. Naunyn-Schmoedeberg's Arch. Pharmacol. **351**, 453 (1995).

174. KAPPOCK, T.J., and J.P. CARADONNA: Pterin-Dependent Amino Acid Hydroxylases. Chem. Rev. **96**, 2659 (1996).

175. HUFTON, S.E., I.G. JENNINGS, and R.G.H. COTTON: Structure and Function of the Aromatic Amino Acid Hydroxylases. Biochem. J. **311**, 353 (1995).

176. SONO, M., M.P. ROACH, E.D. COULTER, and J.H. DAWSON: Heme-Containing Oxygenases. Chem. Rev. **96**, 2841 (1996).

177. McCALL, T.B., N.K. BOUGHTON-SMITH, R.M.J. PALMER, B.J.R. WHITTLE, and S. MONCADA: Synthesis of Nitric Oxide from L-Arginine by Neutrophils: Release and Interaction with Superoxide Anion. Biochem. J. **261**, 293 (1989).

178. POLLOCK, J.S., U. FÖRSTERMANN, J.A. MITCHELL, T.D. WARNER, H.H.H.W. SCHMIDT, M. NAKANE, and F. MURAD: Purification and Characterization of Particulate Endothelium Derived Relaxing Factor Synthase from Cultured and Native Bovine Aortic Endothelial Cells. Proc. Natl. Acad. Sci. USA **88**, 10480 (1991).

179. SCHMIDT, H.H.H.W., J.S. POLLOCK, M. NAKANE, L.D. GORSKY, U. FÖRSTERMANN, and F. MURAD: Purification of a Soluble Isoform of Guanylyl Cyclase-Activating Factor Synthase. Proc. Natl. Acad. Sci. USA **88**, 365 (1991).

180. MIYAHARA, K., T. KAWAMOTO, K. SASE, Y. YUI, K. TODA, L.X. YANG, R. HATTORI, T. AOYAMA, Y. YAMAMOTO, Y. DOI, S. OGOSHI, K. HASHIMOTO, C. KAWAI, S. SASAYAMA, and Y. SHIZUTA: Cloning and Structural Characterization of the Human Endothelial Nitric Oxide Synthase Gene. Eur. J. Biochem. **223**, 719 (1994).

181. HALL, A.V., H. ANTONIOU, Y. WANG, A.H. CHEUNG, A.M. ARBUS, S.L. OLSON, W.C. LU, C.L. KAU, and P.A. MARSDEN: Structural Organization of the Human Neuronal Nitric Oxide Synthase Gene (NOS1). J. Biol. Chem. **269**, 33082 (1994).

182. BREDT, D.S., C.D. FERRIS, and S.H. SNYDER: Nitric Oxide Synthase Regulatory Sites: Phosphorylation by Cyclic AMP-Dependent Protein Kinase, Protein Kinase C, and Calcium Calmodulin Protein Kinase. Identification of Flavin and Calmodulin Binding Sites. J. Biol. Chem. **267**, 10976 (1992).

183. ABU-SOUD, H.M., and D.J. STUEHR: Nitric Oxide Synthases Reveal a Role For Calmodulin in Controlling Electron Transfer. Proc. Natl. Acad. Sci. USA **90**, 10769 (1993).

184. JAFFREY, S.R., and S.H. SNYDER: PIN: an Associated Protein Inhibitor of Neuronal Nitric Oxide Synthase. Science **274**, 774 (1996).

185. VENEMA, V.J., M.B. MARRERO, and R.C. VENEMA: Bradykinin-Stimulated Protein Tyrosine Phosphorylation Promotes Endothelial Nitric Oxide Synthase Translocation to the Cytoskelton. Biochem. Biophys. Res. Commun. **226**, 703 (1996).

186. VODOVOTZ, Y., D. RUSSELL, Q.W. XIE, C. BOGDAN, and C. MATHAN: Vesicle Membrane Association of Nitric Oxide Synthase in Primary Mouse Macrophages. J. Immunol. **154**, 2914 (1995).

187. DAWSON, T.M., J.P. STEINER, V.L. DAWSON, J.L. DINERMAN, G.R. UHL, and S.H. SNYDER: Immunosuppressant FK506 Enhances Phosphorylation of Nitric Oxide Synthase and Protects Against Glutamate Neurotoxicity. Proc. Natl. Acad. Sci. USA **90**, 9808 (1993).

188. BUSCONI, L., and T. MICHEL: Recombinant Endothelial Nitric Oxide Synthase: Posttranslational Modification in a Baculovirus Expression System. Molec. Pharmac. **47**, 655 (1995).

189. HIRATA, K., R. KURODA, T. SAKODA, M. KATAYAMA, N. INOUE, M. SUEMATSU, S. KAWASHIMA, and M. YOKOYAMA: Inhibition of Endothelial Nitric Oxide Synthase Activity by Protein Kinase C. Hypertension **25**, 180 (1995).

190. GARCÍA-CARDENA, G., R. FAN, D.F. STERN, J.W. LIU, and W.C. SESSA: Endothelial Nitric Oxide Synthase Is Regulated by Tyrosine Phosphorylation and Interacts with Caveolin-1. J. Biol. Chem. **271**, 27237 (1996).

191. MATSUBARA, M., K. TITANI, and H. TANIGUCHI: Interaction of Calmodulin-Binding Domain Peptides of Nitric Oxide Synthase with Membrane Phospholipids: Regulation by Protein Phosphorylation and Ca^{2+}-Calmodulin. Biochemistry **35**, 14651 (1996).

192. MICHEL, T., G.K. LI, and L. BUSCONI: Phosphrylation and Subcellular Translocation of Endothelial Nitric Oxide Synthase. Proc. Natl. Acad. Sci. USA **90**, 6252 (1993).

193. ROBINSON, L.J., L. BUSCONI, and T. MICHEL: Agonist-Modulated Palmitoylation of Endothelial Nitric Oxide Synthase. J. Biol. Chem. **270**, 995 (1995).

194. MACALLISTER, R.J., H. PARRY, M. KIMOTO, T. OGAWA, R.J. RUSSELL, H. HODSON, G.S. WHITLEY, and P. VALLANCE: Regulation of Nitric Oxide Synthesis by Dimethylarginine Dimethylaminohydrolase. Br. J. Pharmac. **119**, 1533 (1996).

195. MACALLISTER, R.J., M.H. RAMBAUSEK, P. VALLANCE, D. WILLIAMS, K. H. HOFFMANN, and E. RITZ: Concentration of Dimethyl-L-arginine in the Plasma of Patients with End Stage Renal Failure Nephrology Dialysis Transplantation **11**, 2449 (1996).

196. CHO, H.J., Q.W. XIE, J. CALAYCAY, R.A. MUMFORD, K.M. SWIDEREK, T.D. LEE, and C. NATHAN: Calmodulin Is a Subunit of Nitric Oxide Synthase from Macrophages. J. Exp. Med. **176**, 599 (1992).

197. NATHAN, C., and Q.W. XIE: Regulation of Biosynthesis of Nitric Oxide. J. Biol. Chem. **269**, 13725 (1994).

198. KRÖNCKE, K.D., K. FEHSEL, and V. KOLBBACHOFEN: Inducible Nitric Oxide Synthase and Its Product Nitric Oxide, a Small Molecule with Complex Biological Activities. Biol. Chem. **376**, 327 (1995).

199. CLOSS, E.I.: CATs, a Family of 3 Distinct Mammalian Cationic Amino Acid Transporters. Amino Acids **11**, 193 (1996).

200. STEVENS, B.R., D.K. KAKUDA, K. YU, M. WATERS, C.B. VO, and M.K. RAIZADA: Induced Nitric Oxide Synthesis Is Dependent on Induced Alternatively Spliced CAT-2

Encoding-L-Arginine Transport in Brain Astrocytes. J. Biol. Chem. **271**, 24017 (1996).

201. Closs, E.I., F.Z. Basha, A. Habermeier, and U. Förstermann: Interference of L-Arginine Analogues with L-Arginine Transport Mediated by the Y^+ Carrier hCAT–2B. Nitric Oxide **1**, 65 (1997).

202. Sonoki, T., A. Nagasaki, T. Gotoh, M. Takiguchi, M. Takeya, H. Matsuzaki, and M. Mori: Coinduction of Nitric Oxide Synthase and Arginase I in Cultured Rat Peritoneal Macrophages and Rat Tissues *in vivo* by Lipopolysaccharide. J. Biol. Chem. **272**, 3689 (1997).

203. Vockley, J.G., C.P. Jenkinson, H. Shukla, R.M. Kern, W.W. Grody, and S.D. Cederbaum: Cloning and Characterization of the Human Type II Arginase Gene. Genomics **38**, 118 (1996).

204. Simmons, W.W., D. Ungureanu-Longrois, G.K. Smith, T.W. Smith, and R.A. Kelly: Glucocorticoids Regulate Inducible Nitric Oxide Synthase by Inhibiting Tetrahydrobiopterin Synthesis and L-Arginine Transport. J. Biol. Chem. **271**, 23928 (1996).

205. Nathan, C.: Nitric Oxide as a Secretory Product of Mammalian Cells. FASEB J. **6**, 3051 (1992).

206. Lee, K.H., M.Y. Baek, K.Y. Moon, W.K. Song, C.H. Chung, D.B. Ha, and M.S. Kang: Nitric Oxide as a Messenger Molecule for Myoblast Fusion. J. Biol. Chem. **269**, 14371 (1994).

207. Lin, A.W., C.C. Chang, and C.C. McCormick: Molecular Cloning and Expression of an Avian Macrophage Nitric Oxide Synthase cDNA and the Analysis of the Genomic 5'-Flanking Region. J. Biol. Chem. **271**, 11911 (1996).

208. Hylland, P., and G.E. Nilsson: Evidence that Acetylcholine Mediates Increased Cerebral Blood Flow Velocity in Crucian Carp through a Nitric Oxide-Dependent Mechanism. J. Cerebral Blood Flow and Metabolism **15**, 519 (1995).

209. Yin, Z., T.J. Lam, and Y.M. Sin: Cytokine-Mediated Antimircobial Immune Response of Catfish, *Clarias gariepinus*, as a Defence Against *Aeromonas hydrophila*. Fish & Shellfish Immunology **7**, 93 (1997).

210. Mustafa, T., C. Agnisola, and J.K. Hansen: Evidence for NO-Dependent Vasodilation in the Trout (*Oncorhynchus mykiss*) Coronary System. J. Comp. Physiol. **167**, 98 (1997).

211. Holmqvist, B., and P. Ekstrom: Subcellular Localization of Neuronal Nitric Oxide Synthase in the Brain of a Teleost; An Immunoelectron and Confocal Microscopical Study. Brain Res. **745**, 67 (1997).

212. Bell, T.D., A.E. Pereda, and D.S. Faber: Nitric Oxide Synthase Distribution in the Goldfish Mauthner Cell. Neuroscience Letters **226**, 187 (1997).

213. Radomski, M.W., J.F. Martin, and S. Moncada: Oxide Synthesis of Nitric Oxide by the Hemocytes of the American Horseshoe Crab (*Limulus Polyphemus*). Phil. Trans. R. Soc. (B) **334**, 129 (1991).

214. Ribeiro, J.M.C., and R.H. Nussenzveig: Nitric Synthase Activity from a Hematophagous Insect Salivary Gland. FEBS Letters **330**, 165 (1993).

215. Regulski, M., and T. Tully: Molecular and Biochemical Characterization of dNOS: A Drosophila Ca^{2+}/Calmodulin-Dependent Nitric Oxide Synthase. Proc. Natl. Acad. Sci. USA **92**, 9072 (1995).

216. Martinez, A.: Nitric Oxide Synthase in Invertebrates. Histochem. J. **27**, 770 (1995).

217. Johansson, K.U.I., and M. Carlberg: NO Synthase – What Can Research on Ivertebrates Add to What Is Already Known. Adv. Neuroimmunol. **5**, 431 (1995).

218. Jacklet, J.W., and M. Gruhn: Nitric Oxide as a Putative Transmitter in Aplysia: Neural Circuits and Membrane Effects. Netherlands J. Zool. **44**, 524 (1994).

219. JACKLET J.W., and M. GRUHN: Colocalization of NADPH-Diaphorase and Myomodulin in Synaptic Glomeruli of Aplysia. Neuroreport **5**, 1841 (1994).
220. JACKLET, J.W.: Nitric Oxide Is Used as an Orthograde Cotransmitter at Identified Histaminergic Synapses. J. Neurophysiol. **74**, 891 (1995).
221. ELPHICK, M.R., R.C. RAYNE, V. RIVEROS-MORENO, S. MONCADA, and M. OSHEA: Nitric Oxide Synthesis in Locust Olfactory Interneurons. J. Exp. Biol. **198**, 821 (1995).
222. ELPHICK, M.R., G. KEMENES, K. STARAS, and M. OSHEA: Behavioral Role for Nitric Oxide in Chemosensory Activation of Feeding in a Mollusk. J. Neurosci. **15**, 7653 (1995).
223. ELPHICK, M.R., L. WILLIAMS, and M. OSHEA: New Features of the Locust Optic Lobe: Evidence of a Role for Nitric Oxide in Insect Vision. J. Exp. Biol. **199**, 2395 (1996).
224. LIN, M.F., and E.M. LEISE: NADPH-Diaphorase Activity Changes during Gangliogenesis and Metamorphosis in the Gastropod Mollusk *Ilyanassa Obsoleta*. J. Comp. Neurol. **374**, 194 (1996).
225. MOROZ, L.L., and R. GILLETTE: From *Polyplacophora* to *Cephalopoda* – Comparative Analysis of Nitric Oxide Signaling in Mollusca. Acta Biologica Hungarica **46**, 169 (1995).
226. MULLER, U., and G. BICKER: Calcium-Activated Release of Nitric Oxide and Cellular Distribution of Nitric Synthesizing Neurons in the Nervous System of the Locust. J. Neurosci. **14**, 7521 (1994).
227. MULLER, U.: The Nitric Oxide System in Insects. Progress in Neurobiology **51**, 363 (1997).
228. NINNEMANN, H., and J. MAIER: Indications for the Occurrence of Nitric Oxide Synthases in Fungi and Plants and the Involvement in Photoconidiation of *Neurospora crassa*. Photochem. Photobiol. **64**, 393 (1996).
229. LESHEM, Y.Y.: Nitric Oxide in Biological Systems. Plant Growth Regulation **18**, 155 (1996).
230. SEN, S., and I.R. CHEEMA: Nitric Oxide Synthase and Calmodulin Immunoreactivity in Plant Embryonic Tissue. Biochem. Arch. **11**, 221 (1995).
231. CUETO, M., O. HERNANDEZ-PERERA, R. MARTIN, M.L. BENTURA, J. RODRIGO, S. LAMAS, and M.P. GOLVANO: Presence of Nitric Oxide Synthase Activity in Roots and Nodules of *Lupinus Albus*. FEBS Letters **398**, 159 (1996).
232. CHEN, Y.J., and J.P.N. ROSAZZA: A Bacterial Nitric Oxide Synthase from a Nocardia Species. Biochem. Biophys. Res. Commun. **203**, 1251 (1994).
233. KWIATKOWSKI, A.V., and J.P. SHAPLEIGH: Requirement of Nitric Oxide for Induction of Genes Whose Products are Involved in Nitric Oxide Metabolism in *Rhodobacter Sphaeroides 2.4.3*. J. Biol. Chem. **271**, 24382 (1996).
234. HOOPER, A.B.: Ammonia Oxidation and Energy Transduction in the Nitrifying Bacteria. Microbial Chemoautotrophy, 133 (1984).
235. FÖRSTERMANN, U., M. NAKANE, W.R. TRACEY, and J.S. POLLOCK: Isoforms of Nitric Oxide Synthase: Functions in the Cardiovascular System. Eur. Heart J. **14**, 10 (1993).
236. TRACEY, W.R., J.S. POLLOCK, F. MURAD, M. NAKANE, and U. FÖRSTERMANN: Identification of an Endothelial-like Type-III NO Synthase in LLC-PK1 Kidney Epithelial Cells. Am. J. Physiol. **266**, C 22 (1994).
237. BELHASSEN, L., O. FERON, D.M. KAYE, T. MICHEL, and R.A. KELLY: Regulation by cAMP of Post-Translational Processing and Subcellular Targeting of Endothelial Nitric Oxide synthase (Type 3) in Cardiac Myocytes. J. Biol. Chem. **272**, 11198 (1997).
238. SASE, K., and T. MICHEL: Expression and Regulation of Endothelial Nitric Oxide Synthase. Trends in Cardiovascular Medicine **7**, 28 (1997).

239. Feron, O., L. Belhassen, L. Kobzik, T.W. Smith, R.A. Kelly, and T. Michel: Endothelial Nitric Oxide Synthase Targeting to Caveolae: Specific Interactions with Caveolin Isoforms in Cardiac Myocytes and Endothelial Cells. J. Biol. Chem. **271**, 22810 (1996).

240. Odell, T.J., P.L. Huang, T.M. Dawson, J.L. Dinerman, S.H. Snyder, E.R. Kandel, and M.C. Fishman: Endothelial NOS and the Blockade of LTP by NOS Inhibitors in Mice Lacking Neuronal NOS. Science **265**, 542 (1994).

241. Williams, J.H.: Retrograde Messengers and Long Term Potentiation: A Progress Report. J. Lipid Mediators Cell Signalling **14**, 331 (1996).

242. Holscher, C.: Nitric oxide, the Enigmatic Neuronal Messenger: Its Role in Synaptic Plasticity: Trends Neurosci. **20**, 298 (1997).

243. Sessa, W.C., K. Pritchard, N. Seyedi, J. Wang, and T.H. Hintze: Chronic Exercise in Dogs Increases Coronary Vascular Nitric Oxide Production and Endothelial Nitric Oxide Synthase Gene Expression. Circ. Res. **74**, 349 (1994).

244. Awolesi, M.A., W.C. Sessa, and B.E. Sumpio: Cyclic Strain Up-Regulates Nitric Oxide Synthase in Cultured Bovine Aortic Endothelial Cells. J. Clin. Invest. **96**, 1449 (1995).

245. Akgoren, N., M. Fabricius, and M. Lauritzen: Importance of Nitric Oxide for Local Increases of Blood Flow in Rat Cerebellar Cortex During Electrical Stimulation. Proc. Natl. Acad. Sci. USA **91**, 5903 (1994).

246. Garthwaite, J., and C.L. Boulton: Nitric Oxide Signaling in the Central Nervous System. Annu. Rev. Physiol. **57**, 683 (1995).

247. Asano, K., C.B.E. Chee, B. Gaston, C.M. Lilly, C. Gerard, J.M. Drazen, and J.S. Stamler: Constitutive and Inducible Nitric Oxide Synthase Gene Expression, Regulation, and Activity in Human Lung Epithelial Cells. Proc. Natl. Acad. Sci. USA **91**, 10089 (1994).

248. Brenman, J.E., D.S. Chao, H.H. Xia, K. Aldape, and D.S. Bredt: Nitric Oxide Synthase Complexed with Dystrophin and Absent from Skeletal Muscle Sarcolemma in Duchenne Muscular Dystrophy. Cell **82**, 743 (1995).

249. Narayanan, K., and O.W. Griffith: Synthesis of L-Thiocitrulline, L-Homothiocitrulline, and S-Methyl-L-Thiocirtulline – a New Class of Potent Nitric Oxide Synthase Inhibitors. J. Med. Chem. **37**, 885 (1994).

250. Kobzik, L., M.B. Reid, D.S. Bredt, and J.S. Stamler: Nitric Oxide in Skeletal Muscle. Nature **372**, 546 (1994).

251. Wang, T., Z.P. Xie, and B. Lu: Nitric Oxide Mediates Activity-Dependent Synaptic Suppression at Developing Neuromuscular Synapses. Nature **374**, 262 (1995).

252. Yui, Y., R. Hattori, K. Kosuga, H. Eizawa, K. Hiki, and C. Kawai: Purification of Nitric Oxide Synthase from Rat Macrophages. J. Biol. Chem. **266**, 12544 (1991).

253. Kilbourn, R.G., and P. Belloni: Endothelial Cell Production of Nitrogen Oxides in Response to Inteferon γ in Combination with Tumour Necrosis Factor, Inerleukin-1 or Endotoxin. J. Natl. Canc. Inst. **82**, 772 (1990).

254. Gross, S.S., and R. Levi: Tetrahydrobiopterin Synthesis. An Absolute Requirement for Cytokine-Induced Nitric Oxide Generation by Vascular Smooth Muscle. J. Biol. Chem. **267**, 25722 (1992).

255. Schulz, R., E. Nava, and S. Moncada: Induction and Potential Biological Relevance of a Ca^{2+}-Independent Nitric Oxide Synthase in the Myocardium. Br. J. Pharmacol. **105**, 575 (1992).

256. Liu, J., M.L. Zhao, C.F. Brosnan, and S.C. Lee: Expression of Type II Nitric Oxide Synthase in Primary Human Astrocytes and Microglia. Role of II-1 Beta and II-1 Receptor Antagonists. J. Immunol. **157**, 3569 (1996).

257. ARBONES, M.L., J. RIBERA, L. AGULLO, M.A. BALTRONS, A. CASANOVAS, V. RIVEROS MORENO, and A. GARCIA: Characteristics of Nitric Oxide Synthase Type I of Rat Cerebellar Astrocytes. Glia **18**, 224 (1996).

258. NATHAN, C.F., and J.B. HIBBS: Role of Nitric Oxide Synthesis in Macrophage Antimicrobial Activity. Curr. Opin. Immunol. **3**, 65 (1991).

259. KARUPIAH, G., Q.W. XIE, R.M.L. BULLER, C. NATHAN, C. DUARTE, and J.D. MACMICKING: Inhibition of Viral Replication by Interferon-g-Induced Nitric Oxide Synthase. Science **261**, 1445 (1993).

260. VODOVOTZ, Y., M.S. LUCIA, K.C. FLANDERS, L. CHESLER, Q.W. XIE, T.W. SMITH, J. WEIDNER, R. MUMFORD, R. WEBBER, C. NATHAN, A.B. ROBERTS, C.F. LIPPA, and M.B. SPORN: Inducible Nitric Oxide Synthase in Tangle-Bearing Neurons of Patients with Alzheimers Disease. J. Exp. Med. **184**, 1425 (1996).

261. KOBZIK, L., D.S. BREDT, C.J. LOWENSTEIN, J. DRAZEN, B. GASTON, D. SUGARBAKER, and J.S. STAMLER: Nitric Oxide Synthase in Human and Rat Lung: Immunocyto-chemical and Histochemical Localization. Am. J. Respiratory Cell Molec. Biol. **9**, 371 (1993).

262. WOLF, G.: Nitric Oxide and Nitric Oxide Synthase: Biology, Pathology, Localization, Histology and Histopathology **12**, 251 (1997).

263. GUENGERICH, F.P.: Reactions and Significance of Cytochrome P-450 Enzymes. J. Biol. Chem. **266**, 10019 (1991).

264. GUENGERICH, F.P., and T.L. MACDONALD: Chemical Mechanisms of Catalysis by Cytochromes P-450: A Unified View. Acc. Chem. Res. **17**, 9 (1984).

265. GROVES, J.T.: Key Elements of the Chemistry of Cytochrome P-450: the Oxygen Rebound Mechanism. J. Chem. Edu. **62**, 928 (1985).

266. POULOS, T.L., B.C. FINZEL, and A.J. HOWARD: High Resolution Crystal Structure of Cytochrome P450cam. J. Mol. Biol. **195**, 687 (1987).

267. GUNTER, M.J., and P. TURNER: Metalloporphyrins as Models for the Cytochromes P450. Coord. Chem. Rev. **108**, 115 (1991).

268. HASEMANN, C.A., K.G. RAVICHANDRAN, J.A. PETERSON, and J. DEISENHOFER: Crystal Structure and Refinement of Cytochrome $P450_{TERP}$ at 2.3 Å Resolution. J. Mol. Biol. **236**, 1169 (1994).

269. RAVICHANDRAN, K.G., S.S. BODDUPALLI, C.A. HASEMANN, J.A. PETERSON, and J. DEISENHOFER: Crystal Structure of Hemoprotein Domain of $P450_{BM-3}$, a Prototype for Microsomal P450s. Science **261**, 731 (1993).

270. CUPP-VICKERY, J., and T.L. POULOS: Structure of Cytochrome $P450_{EryF}$ involved in Erythromycin Biosynthesis. Nat. Struct. Biol. **2**, 144 (1995).

271. HARRIS, D.L., and G.H. LOEW: A Role for Thr 252 in Cytochrome P450*cam* Oxygen Activation. J. Am. Chem. Soc. **116**, 11671 (1994).

272. ULLRICH, V., and H.J. STAUDINGER: Studies on Oxygen Activation in Model Systems for Microsomal Mixed Function Oxidases. In: Biological and Chemical Aspects of Monooxygenases, 235 (1966).

273. OSTOVÍC, D., G.-X. HE, and T.C. BRUICE: Mechanisms of Reaction of Hypervalent Oxochromium, Iron, and Manganese Tetraphenylporphyrins with Alkenes. In: Metal-loporphyrins in Catalytic Oxidations, Chapter 2 (1994).

274. DAWSON, J.H., and M. SONO: Cytochrome P450 and Chloroperoxidase: Thiolate-Ligated Heme Enzymes. Spectroscopic Determination of Their Active Site Structures and Mechanistic Implications of Thiolate Ligation. Chem. Rev. **87**, 1255 (1987).

275. NEWCOMB, M., M.-H. LE TADIC-BIADATTI, D.L. CHESTNEY, E.S. ROBERTS, and P.F. HOLLENBERG: A Nonsynchronous Concerted Mechanism For Cytochrome P450 Catalyzed Hydroxylation. J. Am. Chem. Soc. **117**, 12085 (1995).

276. Wilkinson, B., M. Zhu, N.D. Priestley, H.H.T. Nguyen, H. Morimoto, P.G. Williams, S.I. Chan, and H.G. Floss: A Concerted Mechanism for Ethane Hydroxylation by the Particulate Methane Monooxygenase from *Methylococcus capsulatus* (Bath). J. Am. Chem. Soc. **118**, 921 (1996).

277. Shestakov, A.F., and A.E. Shilov: 5-Coordinate Carbon Hydroxylation Mechanism. Journal of Molecular Catalysis A-Chemical **105**, 1 (1996).

278. Ebner, T., C.O. Meese, and M. Eichelbaum: Mechanism of Cytochrome P450 2D6-Catalyzed Sparteine Metabolism in Humans. Molec. Pharmac. **48**, 1078 (1995).

279. MacDonald, T.L., W.G. Gutheim, R.B. Martin, and F.P. Guengerich: Oxidation of Substituted N,N-Dimethylanilines by Cytochrome P-450: Estimation of the Effective Oxidation-Reduction Potential of Cytochrome P-450. Biochemistry **28**, 2071 (1989).

280. Grace, J.M., M.T. Kinter, and T.L. MacDonald: Atypical Metabolism of Deprenyl and Its Enantiomer, (S)-$(+)$-N,α-Dimethyl-N-propynylphenethylamine by Cytochrome P450 2D6. Chem. Res. Toxicol. **7**, 286 (1994).

281. Eberson, L.: Calculation of the P-450 Redox Potential from Kinetic Data by the Marcus Treatment: Feasible or Not? Acta. Chem. Scand. **44**, 733 (1990).

282. Carlson, T.J., J.P. Jones, L. Peterson, N. Castagnoli, K.R. Iyer, and W.F. Trager: Stereoselectivity and Isotope Effects Associated with Cytochrome P450-Catalyzed Oxidation of (S)-Nicotine: The Possibility of Initial Hydrogen Atom Abstraction in the Formation of the Delta-1,5-Nicotinium Ion. Drug Metab. Disp. **23**, 749 (1995).

283. Nelsen, S.F., and J.T. Ippoliti: On the Deprotonation of Trialkylamine Cation Radicals by Amines. J. Am. Chem. Soc. **108**, 4879 (1986).

284. Dinnocenzo, J.P., and T.E. Banach: Deprotonation of Tertiary Amine Cation Radicals: A Direct Experimental Approach. J. Am. Chem. Soc. **111**, 8646 (1989).

285. Parker, V.D., and M. Tilset: Facile Proton Transfer Reactions of N,N-Dimethylaniline Cation Radicals. J. Am. Chem. Soc. **113**, 8778 (1991).

286. Okazaki, O., and F.P. Guengerich: Evidence for Specific Base Catalysis in N-Dealkylation Reactions Catalyzed by Cytochrome P450 and Chloroperoxidase. J. Biol. Chem. **268**, 1546 (1993).

287. Guengerich, F.P., L.C. Bell, and O. Okazaki: Interpretations of Cytochrome P450 Mechanisms from Kinetic Studies. Biochimie **77**, 573 (1995).

288. Guengerich, F.P., O. Okazaki, Y. Seto, and T.L. MacDonald: Radical Cation Intermediates in N-Dealkylation Reactions. Xenobiotica **25**, 689 (1995).

289. Dinnocenzo, J.P., S.B. Karki, and J.P. Jones: On Isotope Effects for the Cytochrome P-450 Oxidation of Substituted N,N-Dimethylanilines. J. Am. Chem. Soc. **115**, 7111 (1993).

290. Karki, S.B., J.P. Dinnocenzo, J.P. Jones, and K.R. Korzekwa: Mechanism of Oxidative Amin Dealkylation of Substituted N,N-Dimethylanilines by Cytochrome P-450: Application of Isotope Effects. J. Am. Chem. Soc. **117**, 3657 (1995).

291. Karki, S.B., and J.P. Dinnocenzo: On the Mechanism of Amine Oxidations by P450. Xenobiotica **25**, 711 (1995).

292. Guengerich, F.P., C.H. Yun, and T.L. MacDonald: Evidence for a 1-Electron Oxidation Mechanism in N-Dealkylation of N,N-Dialkylanilines by Cytochrome P450 2B1: Kinetic Hydrogen Isotope Effects, Linear Free Energy Relationships, Comparisons with Horseradish Peroxidase, and Studies with Oxygen Surrogates. J. Biol. Chem. **271**, 27321 (1996).

293. Skiles, G.L., and G.S. Yost: Mechanistic Studies on the Cytochrome P450-Catalyzed Dehydrogenation of 3-Methylindole. Chem. Res. Toxicol. **9**, 291 (1996).

294. Wolff, T., L.M. Distlerath, M.T. Worthington, J.D. Groopman, G.J. Hammons, F.F. Kadlubar, R.A. Prough, M.V. Martin, and F.P. Guengerich: Substrate Specificity of

Human Liver Cytochrome P-450 Debrisoquine 4-Hydroxylase Probed Using Immunochemical Inhibition and Chemical Modeling. Cancer Res. **45**, 2116 (1985).

295. ISLAM, S.A., C.R. WOLF, M.S. LENNARD, and M.J.E. STERNBERG: A Three-Dimensional Molecular Template for Substrates of Human Cytochrome P450 Involved in Debrisoquine 4-Hydroxylation. Carcinogenesis **12**, 2211 (1991).

296. KOYMANS, L., N.P.E. VERMEULEN, S.A.B.E. VAN ACKER, J.M. TE KOPPELE, J.J.P. HEYKANTS, K. LAVRIJSEN, W. MEULDERMANS, and G.M. DONNÉ-OP DEB KELDER: A Predictive Model for Substrates of Cytochrome P450-Debrisoquine (2D6). Chem. Res. Toxicol. **5**, 211 (1992).

297. STROBL, G.R., S. KRUEDENER, J. STÖCKIGT, F.P. GUENGERICH, and T. WOLFF: Development of a Pharmacophore for Inhibition of Human Liver Cytochrome P-450 2D6: Molecular Modeling and Inhibition Studies. J. Med. Chem. **36**, 1136 (1993).

298. CHEN, P.-F., A.-L. TSAI, V. BERKA, and K.K. WU: Mutation of Glu-361 in Human Endothelial Nitric Oxide Synthase Selectively Abolishes L-Arginine Binding without Perturbing the Behavior of Heme and Other Redox Centers. J. Biol. Chem. **272**, 6114 (1997).

299. GACHHUI, R., D.K. GHOSH, C.Q. WU, J. PARKINSON, B.R. CRANE, and D.J. STUEHR: Mutagenesis of Acidic Residues in the Oxygenase Domain of Inducible Nitric Oxide Synthase Identifies a Glutamate Involved in Arginine Binding. Biochemistry **36**, 5097 (1997).

300. AKHTAR, M., and J.N. WRIGHT: A Unified Mechanistic View of Oxidative Reactions Catalyzed by P-450 and Related Fe-Containing Enzymes. Nat. Prod. Rep. **8**, 527 (1991).

301. AKHTAR, M., V.C.O. NJAR, and J.N. WRIGHT: Mechanistic Studies on Aromatase and Related C–C Bond Cleaving P-450 Enzymes. J. Steroid Biochem. Mol. Biol. **44**, 375 (1993).

302. AKHTAR, M., M.R. CALDER, D.L. CORINA, and J.N. WRIGHT: Mechanistic Studies on C-19 Demethylation in Estrogen Biosynthesis. Biochem. J. **201**, 569 (1982).

303. VAZ, A.D.N., E.S. ROBERTS, and M.J. COON: Olefin Formation in the Oxidative Deformylation of Aldehydes by Cytochrome P-450: Mechanistic Implications for Catalysis by Oxygen-Derived Peroxide. J. Am. Chem. Soc. **113**, 5886 (1991).

304. COLE, P.A., and C.H. ROBINSON: Mechanistic Studies on a Placental Aromatase Model Reaction. J. Am. Chem. Soc. **113**, 8130 (1991).

305. ROBERTS, E.S., A.D.N. VAZ, and M.J. COON: Catalysis by Cytochrome P-450 of an Oxidative Reaction in Xenobiotic Aldehyde Metabolism: Deformylation with Olefin Formation. Proc. Natl. Acad. Sci. USA **88**, 8963 (1991).

306. COLLMAN, J.P., T.R. HALPERT, and K.S. SUSLICK: Dioxygen Binding to Heme Proteins and Their Synthetic Analogs. In: Metal Ion Activation of Dioxygen, 1 (1980).

307. KORTH, H.G., R. SUSTMANN, C. THATER, A.R. BUTLER, and K.U. INGOLD: On the Mechanism of the Nitric Oxide Synthase-catalysed Conversion of N^ω-Hydroxy-L-arginine to Citrulline and Nitric Oxide. J. Biol. Chem. **269**, 17776 (1994).

308. KERWIN, J.F., J.R. LANCASTER, and P.L. FELDMAN: Nitric Oxide: A New Paradigm for Second Messengers. J. Med. Chem. **38**, 4343 (1995).

309. TSAI, A.L., V. BERKA, P.F. CHEN, and G. PALMER: Characterization of Endothelial Nitric Oxide Synthase and Its Reaction with Ligand by Electron Paramagnetic Resonance Spectroscopy. J. Biol. Chem. **271**, 32563 (1996).

310. WASIELEWSKA, E., M. WITKO, G. STOCHEL, and Z. STASICKA: Biosynthesis of Nitric Oxide – Quantum Chemical Modelling of N^ω-Hydroxy-L-arginine Formation. Chemistry – a European Journal **3**, 609 (1997).

311. PUFAHL, R.A., J.S. WISHNOK, and M.A. MARLETTA: Hydrogen Peroxide-Supported Oxidation of N^G-hydroxy-L-arginine by Nitric Oxide Synthase. Biochemistry **34**, 1930 (1995).

312. ABU-SOUD, H.M., P.L. FELDMAN, P. CLARK, and D.J. STUEHR: Electron Transfer in the Nitric Oxide Synthases: Characterization of L-Arginine Analogs that Block Heme Iron Reduction. J. Biol. Chem. **269**, 32318 (1994).

313. PRESTA, A., J. LIU, W.C. SESSA, and D.J. STUEHR: Substrate Binding and Calmodulin Binding to Endothelial Nitric Oxide Synthase Coregulate Its Enzymatic Activity. Nitric Oxide: Chemistry and Biology **1**, 74 (1997).

314. MARLETTA, M.A., P.S. YOON, R. IYENGAR, C.D. LEAF, and J.S. WISHNOK: Macrophage Oxidation of L-Arginine to Nitrite and Nitrate: Nitric Oxide Is an Intermediate. Biochemistry **27**, 8706 (1988).

315. MARLETTA, M.A.: Nitric Oxide: Biosynthesis and Biological Significance. TIBS **14**, 488 (1989).

316. PRINCE, R.C., and G.N. GEORGE: The Remarkable Complexity of Hydroxylamine Oxidoreductase. Nature Struct. Biol. **4**, 247 (1997).

317. DEMASTER, E.G., L. RAIJ, S.L. ARCHER, and E.K. WEIR: Hydroxylamine Is a Vasorelaxant and a Possible Intermediate in the Oxidative Conversion of L-Arginine to Nitric Oxide. Biochem. Biophys. Res. Commun. **163**, 527 (1989).

318. SENNEQUIER, N., J.L. BOUCHER, P. BATTIONI, and D. MANSUY: Superoxide Anion Efficiently Performs the Oxidative Cleavage of C=NOH Bonds of Amidoximes and N-Hydroxyguanidines with Formation of Nitrogen Oxides. Tetrahedron Lett. **36**, 6059 (1995).

319. EVERETT, S.A., M.F. DENNIS, K.B. PATEL, M.R.L. STRATFORD, and P. WARDMAN: Oxidative Denitrification of N^{ω}-Hydroxy-L-arginine by the Superoxide Radical Anion. Biochem. J. **317**, 17 (1996).

320. VETROVSKY, P., J.C. STOCLET, and G. ENTLICHER: Possible Mechanism of Nitric Oxide Production from N^{G}-Hydroxy-L-arginine or Hydroxylamine by Superoxide Ion. Int. J. Biochem. Cell Biol. **28**, 1311 (1996).

321. MODOLELL, M., K. EICKMANN, and G. SOLER: Oxidation of N^{G}-Hydroxy-L-Arginine to Nitric Oxide Mediated by Respiratory Burst: An Alternative Pathway to NO Synthesis. FEBS Letters **401**, 123 (1997).

322. KWON, N.S., C.F. NATHAN, C. GILKER, O.W. GRIFFITH, D.E. MATTHEWS, and D.J. STUEHR: L-Citrulline Production from L-Arginine by Macrophage Nitric Oxide Synthase. The Ureido Oxygen Derives from Dioxygen. J. Biol. Chem. **265**, 13442 (1990).

323. STUEHR, D.J., and O.W. GRIFFITH: Mammalian Nitric Oxide Synthases. Adv. Enzymol. **65**, 287 (1992).

324. SAPSE, A.M., L. HERZIG, and G. SNYDER: A Self-Consistent Field Molecular Orbital Study of Hydroxyguanidine. Cancer Res. **41**, 1824 (1981).

325. STUEHR, D.J., A.M. SAPSE, and D.S. SAPSE: An *Ab Initio* Study of an Oxidative Mechanism that Forms Nitric Oxide from the N-Hydroxyguanidinium Ion. Struct. Chem. **4**, 143 (1993).

326. STUEHR, D.J., N.S. KWON, C.F. NATHAN, O.W. GRIFFITH, P.L. FELDMAN, and J. WISEMAN: N^{ω}-Hydroxy-L-arginine Is an Intermediate in the Biosynthesis of Nitric Oxide from L-Arginine. J. Biol. Chem. **266**, 6259 (1991).

327. FELDMAN, P.L., O.W. GRIFFITH, H. HONG, and D.J. STUEHR: Irreversible Inactivation of Macrophage and Brain Nitric Oxide Synthase by L-N^{G}-Methylarginine Requires NADPH-Dependent Hydroxylation. J. Med. Chem. **36**, 491 (1993).

328. ANDRONIK-LION, V., J.L. BOUCHER, M. DELAFORGE, Y. HENRY, and D. MANSUY: Formation of Nitric Oxide by Cytochrome P450-Catalysed Oxidation of Aromatic Amidoximes. Biochem. Biophys. Res. Commun. **185**, 452 (1992).

329. BOUCHER, J.L., A. GENET, S. VADON, M. DELAFORGE, and D. MANSUY: Formation of Nitrogen Oxides and Citrulline Upon Oxidation of N^{ω}-Hydroxy-L-arginine. Biochem. Biophys. Res. Commun. **184**, 1158 (1992).

330. RENAUD, J.P., J.L. BOUCHER, S. VADON, M. DELAFORGE, and D. MANSUY: Particular Ability of Liver P450s3a to Catalyze the Oxidation of N^{ω}-Hydroxyarginine to Citrulline and Nitrogen Oxides and Occurrence in NO Synthases of a Sequence Very Similar to the Heme Binding Sequence in P450s. Biochem. Biophys. Res. Commun. **192**, 53 (1993).

331. BURSTYN, J.N., J.A. ROE, A.R. MIKSZTAL, B.A. SHAEVITZ, G. LANG, and J.S. VALENTINE: Magnetic and Spectroscopic Characterisation of an Iron Porphyrin Peroxide Complex: Peroxooctaethylporphyrin⁻. J. Am. Chem. Soc. **110**, 1382 (1988).

332. CLEMENT, B., M.H. SCHULTZE-MOSGAU, and H. WOHLERS: Cytochrome P450 Dependent N-Hydroxylation of a Guanidine (Debrisoquine), Microsomal Catalyzed Reduction and Further Oxidation of the N-Hydroxy-guanidine Metabolite to the Urea Derivative: Similarity with the Oxidation of Arginine to Citrulline and Nitric Oxide. Biochem. Pharmac. **46**, 2249 (1993).

333. CLEMENT, B., M.H. SCHULTZE-MOSGAU, P.H. RICHTER, and A. BESCH: Cytochrome P450-Dependent N-Hydroxylation of an Aminoguanidine (Amidinohydrazone) and Microsomal Retroreduction of the N-Hydroxylated Product. Xenobiotica **24**, 671 (1994).

334. PUFAHL, R.A., and M.A. MARLETTA: Oxidation of N^{G}-Hydroxy-L-arginine by Nitric Oxide Synthase: Evidence for the Involvement of the Heme in Catalysis. Biochem. Biophys. Res. Commun. **193**, 963 (1993).

335. CAMPOS, K.L., J. GIOVANELLI, and S. KAUFMAN: Characteristics of the Nitric Oxide Synthase-Catalyzed Conversion of Arginine to N-Hydroxyarginine, the First Oxygenation Step in the Enzymatic-Synthesis of Nitric Oxide. J. Biol. Chem. **270**, 1721 (1995).

336. FELDMAN, P.L., O.W. GRIFFITH, and D.J. STUEHR: The Surprising Life of Nitric Oxide. Chem. Eng. News **71**, 26 (1993).

337. BAILEY, D.M., and C.G. DEGRAZIA: Hydroxyguanidines. A New Class of Antihypertensive Agents. J. Med. Chem. **16**, 151 (1973).

338. ISHIKAWA, T., M. IKENO, T. SAKAMAKI, K. SATO, and K. HIGUCHI: Photo-Sensitized Oxygenation of Phenethylguanidoxime – a Possible Chemical Model For the Biological Oxidation of N^{ω}-hydroxy-L- Arginine to L-Citrulline. Tetrahedron Lett. **37**, 4393 (1996).

339. VAZ, A.D.N., S.J. PERNECKY, G.M. RANER, and M.J. COON: Peroxo-Iron and Oxenoid-Iron Species as Alternative Oxygenating Agents in Cytochrome P450-Catalyzed Reactions: Switching by Threonine-302 to Alanine Mutagenesis of Cytochrome P450 2B4. Proc. Natl. Acad. Sci. USA **93**, 4644 (1996).

340. SHELDON, R.A.: Oxidation Catalysis by Metalloporphyrins. In: Metalloporphyrins in Catalytic Oxidations, Chapter 12 (1994).

341. PUFAHL, R.A., P.G. NANJAPPAN, R.W. WOODARD, and M.A. MARLETTA: Mechanistic Probes of N-Hydroxylation of L-arginine by the Inducible Nitric Oxide Synthase from Murine Macrophages. Biochemistry **31**, 6822 (1992).

342. OLKEN, N.M., and M.A. MARLETTA: N^{G}-Allyl- and N^{G}-Cyclopropyl-L-arginine: Two Novel Inhibitors of Nitric Oxide Synthase. J. Med. Chem. **35**, 1137 (1992).

343. OLKEN, N.M., and M.A. MARLETTA: N^{G}-Methyl-L-arginine Functions as an Alternate Substrate and Mechanism-Based Inhibitor of Nitric Oxide Synthase. Biochemistry **32**, 9677 (1993).

344. OLKEN, N.M., K.M. RUSCHE, M.K. RICHARDS, and M.A. MARLETTA: Inactivation of Macrophage Nitric Oxide Synthase Activity by N^{G}-Methyl-L-arginine:. Biochem. Biophys. Res. Commun. **177**, 828 (1991).

345. HEINZEL, B., M. JOHN, P. KLATT, E. BOHME, and B. MAYER: Ca^{2+} / Calmodulin-Dependent Formation of Hydrogen Peroxide by Brain Nitric Oxide Synthase. Biochem. J. **281**, 627 (1992).

346. Guengerich, F.P.: Destruction of Heme and Hemoproteins Mediated by Liver Microsomal Reduced Nicotinamide Adenine Nucleotide Phosphate-Cytochrome P-450 Reductase. Biochemistry **17**, 3633 (1978).

347. Schaefer, W.H., T.M. Harris, and F.P. Guengerich: Characterisation of the Enzymatic and Nonenzymatic Peroxidative Degradation of Iron Porphyrins and Cytochrome P-450 Heme. Biochemistry **24**, 3254 (1985).

348. Olken, N.M., Y. Osawa, and M.A. Marletta: Characterization of the Inactivation of Nitric Oxide Synthase by N^G-Methyl-L-arginine: Evidence for Heme Loss. Biochemistry **33**, 14784 (1994).

349. Brewer, C.B., and J.A. Peterson: Single Turnover Kinetics of the Reaction Between Oxycytochrome-P-450cam and Reduced Putidaredoxin. J. Biol. Chem. **263**, 791 (1988).

350. Fukuto, J.M.: Chemistry of N-Hydroxy-L-arginine. Methods Enzymol. **268**, 365 (1996).

351. Vidal, J., S. Damestoy, and A. Collet: Electrophilic Amination Reagents: A New Method for the Preparation of 3-Aryl-N-Boc (or N-Fmoc) Oxaziridines. Tetrahedron Lett. **36**, 1439 (1995).

352. Klatt, P., K. Schmidt, G. Uray, and B. Mayer: Multiple Catalytic Functions of Brain Nitric Oxide Synthase: Biochemical Characterization, Cofactor Requirement, and the Role of $N^{(G.\omega)}$-Hydroxy-L-arginine as an Intermediate. J. Biol. Chem. **268**, 14781 (1993).

353. Klatt, P., K. Schmidt, D. Lehner, O. Glatter, H.P. Bachinger, and B. Mayer: Structural Analysis of Porcine Brain Nitric Oxide Synthase Reveals a Role for Tetrahydrobiopterin and L-Arginine in the Formation of an SDS-Resistant Dimer. EMBO J. **14**, 3687 (1995).

354. Riveros-Moreno, V., B. Heffernan, B. Torres, A. Chubb, I. Charles, and S. Moncada: Purification to Homogeneity and Characterization of Rat Brain Recombinant Nitric Oxide Synthase. Eur. J. Biochem. **230**, 52 (1995).

355. Cho, H.J., E. Martin, Q.W. Xie, S. Sassa, and C. Nathan: Inducible Nitric Oxide Synthase: Identification of Amino Acid Residues Essential for Dimerization and Binding of Tetrahydrobiopterin. Proc. Natl. Acad. Sci. USA **92**, 11514 (1995).

356. Tzeng, E., T.R. Billiar, P.D. Robbins, M. Loftus, and D.J. Stuehr: Expression of Human Inducible Nitric Oxide Synthase in a Tetrahydrobiopterin (H_4B)-Deficient Cell-Line: H_4B Promotes Assembly of Enzyme Subunits into an Active Dimer. Proc. Natl. Acad. Sci. USA **92**, 11771 (1995).

357. Albakri, Q.A., and D.J. Stuehr: Intracellular Assembly of Inducible NO Synthase Is Limited by Nitric Oxide-Mediated Changes in Heme Insertion and Availability. J. Biol. Chem. **271**, 5414 (1996).

358. Xie, Q.W., M. Leung, M. Fuortes, S. Sassa, and C. Nathan: Complementation Analysis of Mutants of Nitric Oxide Synthase Reveals that the Active Site Requires 2 Hemes. Proc. Natl. Acad. Sci. USA **93**, 4891 (1996).

359. Marsden, P.A., K.T. Schappert, H.S. Chen, M. Flowers, C.L. Sundell, J.N. Wilcox, S. Lamas, and T. Michel: Molecular Cloning and Characterization of Human Endothelial Nitric Oxide Synthase. FEBS Letters **307**, 287 (1992).

360. Adachi, H., S. Iida, S. Oguchi, H. Ohshima, H. Suzuki, K. Nagasaki, H. Kawasaki, T. Sugimura, and H. Esumi: Molecular Cloning of a cDNA Encoding an Inducible Calmodulin-Dependent Nitric Oxide Synthase From Rat Liver and Its Expression in COS 1 Cells. Eur. J. Biochem. **217**, 37 (1993)

361. CHARLES, I.G., A. CHUBB, R. GILL, J. CLARE, P.N. LOWE, L.S. HOLMES, M. PAGE, J.G. KEELING, S. MONCADA, and V. RIVEROS-MORENO: Cloning and Expression of a Rat Neuronal Nitric Oxide Synthase Coding Sequence in a Baculovirus Insect Cell System. Biochem. Biophys. Res. Commun. **196**, 1481 (1993).
362. MAIER, R., G. BILBE, J. REDISKE, and M. LOTZ: Inducible Nitric Oxide Synthase from Human Articular Chondrocytes: cDNA Cloning and Analysis of mRNA Expression. Biochimica et Biophysica Acta-Protein Structure and Molecular Enzymology **1208**, 145 (1994).
363. GALEA, E., D.J. REIS, and D.L. FEINSTEIN: Cloning and Expression of Inducible Nitric Oxide Synthase from Rat Astrocytes. Journal Of Neuroscience Research **37**, 406 (1994).
364. GENG, Y.J., M. ALMQVIST, and G.K. HANSSON: cDNA Cloning and Expression of Inducible Nitric Oxide Synthase from Rat Vascular Smooth Muscle Cells. Biochimica et Biophysica Acta-Gene Structure and Expression **1218**, 421 (1994).
365. HOKARI, A., M. ZENIYA, and H. ESUMI: Cloning and Functional Expression of Human Inducible Nitric Oxide Synthase (NOS) cDNA from a Glioblastoma Cell Line A-172. J. Biochem. **116**, 575 (1994).
366. PARK, C.S., K. PARDHASARADHI, C. GIANOTTI, E. VILLEGAS, and G. KRISHNA: Human Retina Expresses Both Constitutive and Inducible Isoforms of Nitric Oxide Synthase mRNA. Biochem. Biophys. Res. Commun. **205**, 85 (1994).
367. IWASHINA, M., Y. HIRATA, T. IMAI, K. SATO, and F. MARUMO: Molecular Cloning of Endothelial, Inducible Nitric Oxide Synthase Gene from Rat Aortic Endothelial Cell. Eur. J. Biochem. **237**, 668 (1996).
368. GARBÁN, H., D. MARQUEZ, T. MAGEE, J. MOODY, T. RAJAVASHISTH, J.A. RODRIGUEZ, A. HUNG, D. VERNET, J. RAJFER, and N.F. GONZÁLEZ-CADAVID: Cloning of Rat an Human Inducible Penile Nitric Oxide Synthase. Application for Gene Therapy of Erectile Dysfunction. Biology of Reproduction **56**, 954 (1997).
369. TSUTSUMISHITA, Y., Y. KAWAI, H. TAKAHARA, T. ONDA, J. MIYOSHI, S. FUTAKI, and M. NIWA: Sequence Analysis of Inducible Nitric Oxide Synthase in Rat Kidney, Lung, and Uterus. Biological & Pharmaceutical Bulletin **19**, 1374 (1996).
370. GNANAPANDITHEN, K., Z.Q. CHEN, C.L. KAU, R.M. GORCZYNSKI, and P.A. MARSDEN: Cloning and Characterization of Murine Endothelial Constitutive Nitric Oxide Synthase. Biochimica et Biophysica Acta-Gene Structure & Expression **1308**, 103 (1996).
371. KARLSEN, A.E., H.U. ANDERSEN, H. VISSING, P.M. LARSEN, S.J. FEY, B.G. CUARTERO, O.D. MADSEN, J.S. PETERSEN, S.B. MORTENSEN, T. MANDRUPPOULSEN, E. BOEL, and J. NERUP: Cloning and Expression of Cytokine-Inducible Nitric Oxide Synthase cDNA From Rat Islets of Langerhans. Diabetes **44**, 753 (1995).
372. ZHANG, J.L., J.M. PATEL, and E.R. BLOCK: Molecular Cloning, Characterization and Expression a Nitric Oxide Synthase from Porcine Pulmonary Artery Endothelial Cells. Comparative Biochemistry and Physiology B-Biochemistry & Molecular Biology **116**, 485 (1997).
373. YUDA, M., M. HIRAI, K. MIURA, H. MATSUMURA, K. ANDO, and Y. CHINZEI: cDNA Cloning, Expression and Characterization of Nitric Oxide Synthase from the Salivary Glands of the Blood Sucking Insect *Rhodnius prolixus*. Eur. J. Biochem. **242**, 807 (1996).
374. BOYHAN, A., D. SMITH, I.G. CHARLES, M. SAQI, and P.N. LOWE: Delineation of the Arginine- and Tetrahydrobiopterin-Binding Sites of Neuronal Nitric Oxide Synthase. Biochem. J. **323**, 131 (1997).

375. BRENMAN, J.E., D.S. CHAO, S.H. GEE, A.W. MCGEE, S.E. CRAVEN, D.R. SANTILLANO, Z.Q. WU, F. HUANG, H.H. XIA, M.F. PETERS, S.C. FROEHNER, and D.S. BREDT: Interaction of Nitric Oxide Synthase with the Postsynaptic Density Protein PSD-95 and a1-Syntrophin Mediated by PDZ Domains. Cell **84**, 757 (1996).

376. PORTER, T.D.: An Unusual Yet Strongly Conserved Flavoprotein Reductase in Bacteria and Mammals. TIBS **16**, 154 (1991).

377. HANIU, M., M.E. MCMANUS, D.J. BIRKETT, T.D. LEE, and J.E. SHIVELY: Structural and Functional Analysis of NADPH-Cytochrome P-450 Reductase From Human Liver: Complete Sequence of Human Enzyme and NADPH-Binding Sites. Biochemistry **28**, 8639 (1989).

378. KARPLUS, P.A., M.J. DANIELS, and J.R. HERRIOTT: Atomic Structure of Ferredoxin-NADP+ Reductase: Prototype for a Structurally Novel Flavoenzyme Family. Science **251**, 60 (1991).

379. SESSA, W.C., C.M. BARBER, and K.R. LYNCH: Mutation of *N*-Myristoylation Site Converts Endothelial Nitric Oxide Synthase from a Membrane to a Cytosolic Protein. Circ. Res. **72**, 921 (1993).

380. LIU, J.W., G. GARCÍA-CARDEÑA, and W.C. SESSA: Biosynthesis and Palmitoylation of Endothelial Nitric Oxide Synthase: Mutagenesis of Palmitoylation Sites, Cysteine-15 and / or Cysteine-26, Argues Against Depalmitoylation-Induced Translocation of the Enzyme. Biochemistry **34**, 12333 (1995).

381. CHEN, P.F., A.L. TSAI, and K.K. WU: Cysteine-184 of Endothelial Nitric Oxide Synthase Is Involved in Heme Coordination and Catalytic Activity. J. Biol. Chem. **269**, 25062 (1994).

382. CHEN, P.F., A.L. TSAI, and K.K. WU: Cysteine-99 of Endothelial, Nitric Oxide Synthase (NOS III) Is Critical for Tetrahydrobiopterin-Dependent NOS III Stability and Activity. Biochem. Biophys. Res. Commun. **215**, 1119 (1995).

383. RICHARDS, M.K., and M.A. MARLETTA: Characterization of Neuronal Nitric Oxide Synthase and a C415H Mutant, Purified from a Baculovirus Overexpression System. Biochemistry **33**, 14723 (1994).

384. RICHARDS, M.K., M.J. CLAGUE, and M.A. MARLETTA: Characterization of C415 Mutants of Neuronal Nitric Oxide Synthase. Biochemistry **35**, 7772 (1996).

385. MCMILLAN, K., and B.S.S. MASTERS: Prokaryotic Expression of the Heme-Binding and Flavin-Binding Domains of Rat Neuronal Nitric Oxide Synthase as Distinct Polypeptides – Identification of the Heme-Binding Proximal Thiolate Ligand as Cysteine-415. Biochemistry **34**, 3686 (1995).

386. CUBBERLEY, R.R., W.K. ALDERTON, A. BOYHAN, I.G. CHARLES, P.N. LOWE, and R.W. OLD: Cysteine-200 of Human Inducible Nitric Oxide Synthase Is Essential for Dimerization of Haem Domains and for Binding of Haem, Nitroarginine and Tetrahydrobiopterin. Biochem. J. **323**, 141 (1997).

387. MAYER, B., E. PITTERS, S. PFEIFFER, W.R. KUKOVETZ, and K. SCHMIDT: A Synthetic Peptide Corresponding to the Putative Dihydrofolate Reductase Domain of Nitric Oxide Synthase Inhibits Uncoupled NADPH Oxidation. Nitric Oxide: Biology and Chemistry **1**, 50 (1997).

388. VENEMA, R.C., H. JU, R. ZOU, J.W. RYAN, and V.J. VENEMA: Subunit Interactions of Endothelial Nitric Oxide synthase: Comparisons to the Neuronal and Inducible Nitric Oxide Synthase Isoforms. J. Biol. Chem. **272**, 1276 (1997).

389. ZHANG, J.L., J.M. PATEL, Y.D. LI, and E.R. BLOCK: Reductase Domain Cysteine 1048 and 1114 Are Critical for Catalytic Activity of Human Endothelial Cell Nitric Oxide Synthase as Probed by Site-Directed Mutagenesis. Biochem. Biophys. Res. Commun. **226**, 293 (1996).

390. Lowe, P.N., D. Smith, D.K. Stammers, V. Riveros-Moreno, S. Moncada, I. Charles, and A. Boyhan: Identification of the Domains of Neuronal Nitric Oxide Synthase by Limited Proteolysis. Biochem. J. **314**, 55 (1996).

391. Sheta, E.A., K. McMillan, and B.S.S. Masters: Evidence for a Bidomain Structure of Constitutive Cerebellar Nitric Oxide Synthase. J. Biol. Chem. **269**, 15147 (1994).

392. Ghosh, D.K., and D.J. Stuehr: Macrophage NO Synthase: Characterization of Isolated Oxygenase and Reductase Domains Reveals a Head-to-Head Subunit Interaction. Biochemistry **34**, 801 (1995).

393. Gachhui, R., A. Presta, D.F. Bentley, H.M. Abusoud, R. Mcarthur, G. Brudvig, D.K. Ghosh, and D.J. Stuehr: Characterization of the Reductase Domain of Rat Neuronal Nitric Oxide Synthase Generated in the Methylotrophic Yeast *Pichia pastoris*: Calmodulin Response Is Complete within the Reductase Domain Itself. J. Biol. Chem. **271**, 20594 (1996).

394. Chen, P.F., A.L. Tsai, V. Berka, and K.K. Wu: Endothelial Nitric Oxide Synthase: Evidence For Bidomain Structure and Successful Reconstitution of Catalytic Activity from Two Separate Domains Generated by a Baculovirus Expression System. J. Biol. Chem. **271**, 14631 (1996).

395. Salerno, J.C., P. Martasek, L.J. Roman, and B.S.S. Masters: Electron Paramagnetic Resonance Spectroscopy of the Heme Domain of Inducible Nitric Oxide Synthase: Binding of Ligands at the Arginine Site Induces Changes in the Heme Ligation Geometry. Biochemistry **35**, 7626 (1996).

396. Stuehr, D.J.: Structure-Function Aspects in the Nitric Oxide Synthases. Annu. Rev. Pharmac. Toxicol. **37**, 339 (1997).

397. Ghosh, D.K., H.M. Abusoud, and D.J. Stuehr: Reconstitution of the 2nd Step in NO Synthesis Using the Isolated Oxygenase and Reductase Domains of Macrophage NO Synthase. Biochemistry **34**, 11316 (1995).

398. Galli, C., R. Macarthur, H.M. Abusoud, P. Clark, D.J. Stuehr, and G.W. Brudvig: EPR Spectroscopic Characterization of Neuronal NO Synthase. Biochemistry **35**, 2804 (1996).

399. Vermilion, J.L., and M.J. Coon: Purified Liver Microsomal NADPH-Cytochrome P-450 Reductase: Spectral Characterization of Oxidation-Reduction States. J. Biol. Chem. **253**, 2694 (1978).

400. Vermilion, J.L., and M.J. Coon: Identification of the High and Low Potential Flavins of Liver Microsomal NADPH-Cytochrome P-450 Reductase. J. Biol. Chem. **253**, 8812 (1978).

401. Yasukochi, Y., J.A. Peterson, and B.S.S. Masters: NADPH-Cytochrome c (P-450) Reductase: Spectrophotometric and Stopped Flow Kinetic Studies on the Formation of Reduced Flavoprotein Intermediates. J. Biol. Chem. **254**, 7097 (1979).

402. Abu-Soud, H.M., L.L. Yoho, and D.J. Stuehr: Calmodulin Controls Neuronal Nitric Oxide Synthase by a Dual Mechanism: Activation of Intradomain and Interdomain Electron Transfer. J. Biol. Chem. **269**, 32047 (1994).

403. Hobbs, A.J., J.M. Fukuto, and L.J. Ignarro: Formation of Free Nitric Oxide from L-Arginine by Nitric Oxide Synthase: Direct Enhancement of Generation by Superoxide Dismutase. Proc. Natl. Acad. Sci. USA **91**, 10992 (1994).

404. Pou, S., W.S. Pou, D.S. Bredt, S.H. Snyder, and G.M. Rosen: Generation of Superoxide by Purified Brain Nitric Oxide Synthase. J. Biol. Chem. **267**, 24173 (1992).

405. Klatt, P., B. Heinzel, M. John, M. Kastner, E. Bohme, and B. Mayer: Ca^{2+}/ Calmodulin-Dependent Cytochrome c Reductase Activity of Brain Nitric Oxide Synthase. J. Biol. Chem. **267**, 11374 (1992).

406. MATSUOKA, A., D.J. STUEHR, J.S. OLSON, P. CLARK, and M. IKEDA-SAITO: L-Arginine and Calmodulin Regulation of the Heme Iron Reactivity in Neuronal Nitric Oxide Synthase. J. Biol. Chem. **269**, 20335 (1994).

407. POZO, D., R.J. REITER, J.R. CALVO, and J.M. GUERRERO: Inhibition of Cerebellar Nitric Oxide Synthase and Cyclic GMP Production by Melatonin *via* Complex Formation with Calmodulin. Journal of Cellular Biochemistry **65**, 430 (1997).

408. PERSECHINI, A., K.J. GANSZ, and R.J. PARESI: Activation of Myosin Light Chain Kinase and Nitric Oxide Synthase Activities by Engineered Calmodulins with Duplicated or Exchanged EF Hand Pairs. Biochemistry **35**, 224 (1996).

409. ZHANG, M.J., T. YUAN, J.M. ARAMINI, and H.J. VOGEL: Interaction of Calmodulin with Its Binding Domain of Rat Cerebellar Nitric Oxide Synthase – a Multinuclear NMR-Study. J. Biol. Chem. **270**, 20901 (1995).

410. VORHERR, T., L. KNOPFEL, F. HOFMANN, S. MOLLNER, T. PFEUFFER, and E. CARAFOLI: The Calmodulin-Binding Domain of Nitric Oxide Synthase and Adenylyl Cyclase. Biochemistry **32**, 6081 (1993).

411. PERSECHINI, A., H.D. WHITE, and K.J. GANSZ: Different Mechanisms for Ca^{2+} Dissociation from Complexes of Calmodulin with Nitric Oxide Synthase or Myosin Light Chain Kinase. J. Biol. Chem. **271**, 62 (1996).

412. ANAGLI, J., F. HOFMANN, M. QUADRONI, T. VORHERR, and E. CARAFOLI: The Calmodulin-Binding Domain of the Inducible (Macrophage) Nitric Oxide Synthase. Eur. J. Biochem. **233**, 701 (1995).

413. STEVENS-TRUSS, R., and M.A. MARLETTA: Interaction of Calmodulin with the Inducible Murine Macrophage Nitric Oxide Synthase. Biochemistry **34**, 15638 (1995).

414. VENEMA, R.C., H.S. SAYEGH, J.D. KENT, and D.G. HARRISON: Identification, Characterization, and Comparison of the Calmodulin-Binding Domains of the Endothelial and Inducible Nitric Oxide Synthases. J. Biol. Chem. **271**, 6435 (1996).

415. ZOCHE, M., M. BIENERT, M. BEYERMANN, and K.W. KOCH: Distinct Molecular Recognition of Calmodulin-Binding Sites in the Neuronal and Macrophage Nitric Oxide Synthases: A Surface Plasmon Resonance Study. Biochemistry **35**, 8742 (1996).

416. WATANABE, Y., Y. HU, and H. HIDAKA: Identification of a Specific Amino Acid Cluster in the Calmodulin-Binding Domain of the Neuronal Nitric Oxide Synthase. FEBS Letters **403**, 75 (1997).

417. ZHANG, M.J., and H.J. VOGEL: Characterization of the Calmodulin-Binding Domain of Rat Cerebellar Nitric Oxide Synthase. J. Biol. Chem. **269**, 981 (1994).

418. PERSECHINI, A., K.J. GANSZ, and R.J. PARESI: A Role in Enzyme Activation for the N-Terminal Leader Sequence in Calmodulin. J. Biol. Chem. **271**, 19279 (1996).

419. PERSECHINI, A., P.M. STEMMER, and I. OHASHI: Localization of Unique Functional Determinants in the Calmodulin Lobes to Individual EF Hands. J. Biol. Chem. **271**, 32217 (1996).

420. SU, Z.Z., M.A. BLAZING, D. FAN, and S.E. GEORGE: The Calmodulin-Nitric Oxide Synthase Interaction: Critical Role of the Calmodulin Latch Domain in Enzyme Activation. J. Biol. Chem. **270**, 29117 (1995).

421. RUAN, J., Q.W. XIE, N. HUTCHINSON, H. CHO, G.C. WOLFE, and C. NATHAN: Inducible Nitric Oxide Synthase Requires Both the Canonical Calmodulin-Binding Domain and Additional Sequences in Order to Bind Calmodulin and Produce Nitric Oxide in the Absence of Free Ca^{2+}. J. Biol. Chem. **271**, 22679 (1996).

422. WU, C.Q., J.G. ZHANG, H. ABU-SOUD, D.K. GHOSH, and D.J. STUEHR: High-Level Expression of Mouse Inducible Nitric Oxide Synthase in *Escherichia Coli* Requires Coexpression with Calmodulin. Biochem. Biophys. Res. Commun. **222**, 439 (1996).

423. COUNTS GERBER, N., and P.R.O. DEMONTELLANO: Active Site Topology of Nitric Oxide Synthase Expressed in *Escherichia coli*. FASEB J. **9**, A1490 (1995).

424. BUSCONI, L., and T. MICHEL: Endothelial Nitric Oxide Synthase: N-Terminal Myristoylation Determines Subcellular Localization. J. Biol. Chem. **268**, 8410 (1993).

425. WOLF, G., S. WURDIG, and G. SCHUNZEL: Nitric Oxide Synthase in Rat Brain Is Predominantly Located at Neuronal Endoplasmic Reticulum: An Electron Microscopic Demonstration of NADPH-Diaphorase Activity. Neuroscience Letters **147**, 63 (1992).

426. HIKI, K., R. HATTORI, C. KAWAI, and Y. YUI: Purification of Insoluble Nitric Oxide Synthase From Rat Cerebellum. J. Biochem. **111**, 556 (1992).

427. HECKER, M., A. MULSCH, and R. BUSSE: Subcellular Localization and Characterization of Neuronal Nitric Oxide Synthase. J. Neurochem. **62**, 1524 (1994).

428. RODRIGO, J., V. RIVEROS-MORENO, M.L. BENTURA, L.O. UTTENTHAL, E.A. HIGGS, A.P. FERNANDEZ, J.M. POLAK, S. MONCADA, and R. MARTINEZ-MURILLO: Subcellular Localization of Nitric Oxide Synthase in the Cerebral Ventricular System, Subfornical Organ, Area Postrema, and Blood Vessels of the Rat Brain. J. Comp. Neurol. **378**, 522 (1997).

429. HIKI, K., Y. YUI, R. HATTORI, H. EIZAWA, K. KOSUGA, and C. KAWAI: Cytosolic and Membrane-Bound Nitric Oxide Synthase. Japanese Journal Of Pharmacology **56**, 217 (1991).

430. DARIUS, S., G. WOLF, P.L. HUANG, and M.C. FISHMAN: Localization of NADPH-Diaphorase / Nitric Oxide Synthase in the Rat Retina: An Electron Microscopic Study. Brain Res. **690**, 231 (1995).

431. ERVASTI, J.M., and K.P. CAMPBELL: Dystrophin and the Membrane Skeleton. Curr. Opin. Cell Biol. **5**, 82 (1993).

432. PONTING, C.P., and C. PHILLIPS: DHR Domains in Syntrophins, Neuronal NO Synthases and other Intracellular Proteins. TIBS **20**, 102 (1995).

433. STRICKER, N.L., K.S. CHRISTOPHERSON, B.A. YI, P.J. SCHATZ, R.W. RAAB, G. DAWES, D.E. BASSETT, D.S. BREDT, and M. LI: PDZ Domain of Neuronal Nitric Oxide Synthase Recognizes Novel C-Terminal Peptide Sequences. Nature Biotechnology **15**, 336 (1997).

434. KENNEDY, M.B.: Origin of PDZ (DHR, GLGF) Domains. Trends Biochem. Sci **20**, 350 (1995).

435. BREDT, D.S.: Targeting Nitric Oxide to Its Targets. Proc. Soc. Exp. Biol. Med. **211**, 41 (1996).

436. PONTING, C.P., C. PHILLIPS, K.E. DAVIES, and D.J. BLAKE: PDZ Domains: Targeting Signalling Molecules to Sub-Membranous Sites. Bioessays **19**, 469 (1997).

437. GOMPERTS, S.N.: Clustering Membrane Proteins: It's All Coming Together with the PSD-95 / SAP90 Protein Family. Cell **84**, 659 (1996).

438. KORNAU, H.C., L.T. SCHENKER, M.B. KENNEDY, and P.H. SEEBURG: Domain Interaction Between NMDA Receptor Subunits and the Postsynaptic Density Protein PSD-95. Science **269**, 1737 (1995).

439. KIM, E., M. NIETHAMMER, A. ROTHSCHILD, Y.N. JAN, and M. SHENG: Clustering of Shaker-Type K+ Channels by Interaction with a Family of Membrane-Associated Guanylate Kinases. Nature **378**, 85 (1995).

440. DOYLE, D.A., A. LEE, J. LEWIS, E. KIM, M. SHENG, and R. MACKINNON: Crystal Structures of a Complexed and Peptide-Free Membrane Protein-Binding Domain: Molecular Basis of Peptide Recognition by PDZ. Cell **85**, 1067 (1996).

441. MORAIS CABRAL, J.H., C. PETOSA, M.J. SUTCLIFFE, S. RAZA, O. BYRON, F. POY, S.M. MARFATIA, A.H. CHISHTI, and R.C. LIDDINGTON: Crystal Structure of a PDZ Domain. Nature **382**, 649 (1996).

442. Vesely, D.L.: Melatonin Enhances Guanylate Cyclase Activity in a Variety of Tissues. Molec. Cell. Biochem. **35**, 55 (1981).

443. Schepens, J., E. Cuppen, B. Wieringa, and W. Hendriks: The Neuronal Nitric Oxide Synthase PDZ Motif Binds to –G(D,E)XV* Carboxyterminal Sequences. FEBS Letters **409**, 53 (1997).

444. Peters, M.F., N.R. Kramarcy, R. Sealock, and S.C. Froehner: b2-Syntrophin: Localization at the Neuromuscular Junction in Skeletal Muscle. Neuroreport **5**, 1577 (1994).

445. Takeuchi, M., Y. Hata, K. Hirao, A. Toyoda, M. Irie, and Y. Takai: Sapaps: A family of PSD-95/SAP90-Associated Proteins Localized at Postsynaptic Density. J. Biol. Chem. **272**, 11943 (1997).

446. Kim, E., S. Naisbitt, Y.P. Hsueh, A. Rao, A. Rothschild, A.M. Craig, and M. Sheng: GKAP, a Novel Synaptic Protein that Interacts with the Guanylate Kinase-Like Domain of the PSD-95/SAP90 Family of Channel Clustering Molecules. Journal of Cell Biology **136**, 669 (1997).

447. Kim, E., K.-O. Cho, A. Rothschild, and M. Sheng: Heteromultimerization and NMDA Receptor Clustering Activity of Chapsyn-110, a Member of the PSD-95 Family of Proteins. Neuron **17**, 103 (1996).

448. Chao, D.S., J.R.M. Gorospe, J.E. Brenman, J.A. Rafael, M.F. Peters, S.C. Froehner, E.P. Hoffman, J.S. Chamberlain, and D.S. Bredt: Selective Loss of Sarcolemmal Nitric Oxide Synthase in Becker Muscular Dystrophy. J. Exp. Med. **184**, 609 (1996).

449. Grozdanovic, Z., G. Gosztonyi, and R. Gossrau: Nitric Oxide Synthase-I (NOS-I) Is Deficient in the Sarcolemma of Striated Muscle Fibers in Patients with Duchenne Muscular Dystrophy, Suggesting an Association With Dystrophin. Acta Histochemica **98**, 61 (1996).

450. Grozdanovic, Z., T. Christova, G. Gosztonyi, H. Mellerowicz, D. Blottner, and R. Gossrau: Absence of Nitric Oxide Synthase I despite the Presence of the Dystrophin Complex in Human Striated Muscle. Histochem. J. **29**, 97 (1997).

451. Chang, W.J., S.T. Iannaccone, K.S. Lau, B.S.S. Masters, T.J. Mccabe, K. McMillan, R.C. Padre, M.J. Spencer, J.G. Tidball, and J.T. Stull: Neuronal Nitric Oxide Synthase and Dystrophin-Deficient Muscular Dystrophy. Proc. Natl. Acad. Sci. USA **93**, 9142 (1996).

452. Li, X.J., A.H. Sharp, S.H. Li, T.M. Dawson, S.H. Snyder, and C.A. Ross: Huntingtin Associated Protein (HAP-1): Discrete Neuronal Localizations in the Brain Resemble Those of Neuronal Nitric Oxide Synthase. Proc. Natl. Acad. Sci. USA **93**, 4839 (1996).

453. Förstermann, U., J.S. Pollock, H.H.H.W. Schmidt, M. Heller, and F. Murad: Calmodulin-Dependent Endothelium-Derived Relaxing Factor Nitric Oxide Synthase Activity Is Present in the Particulate and Cytosolic Fractions of Bovine Aortic Endothelial Cells. Proc. Natl. Acad. Sci. USA **88**, 1788 (1991).

454. Morin, A.M., and A. Stanboli: Nitric Oxide Synthase in Cultured Endothelial Cells of Cerebrovascular Origin: Cytochemistry. Journal of Neuroscience Research **36**, 272 (1993).

455. O'Brien, A.J., H.M. Young, J.M. Povey, and J.B. Furness: Nitric Oxide Synthase Is Localized Predominantly in the Golgi Apparatus and Cytoplasmic Vesicles of Vascular Endothelial Cells. Histochemistry and Cell Biology **103**, 221 (1995).

456. Sessa, W.C., G. García-Cardeña, J.W. Liu, A. Keh, J.S. Pollock, J. Bradley, S. Thiru, I.M. Braverman, and K.M. Desai: The Golgi Association of Endothelial Nitric Oxide Synthase Is Necessary for the Efficient Synthesis of Nitric Oxide. J. Biol. Chem. **270**, 17641 (1995).

457. HECKER, M., A. MULSCH, E. BASSENGE, U. FÖRSTERMANN, and R. BUSSE: Subcellular Localization and Characterization of Nitric Oxide Synthase(s) in Endothelial Cells: Physiological Implications. Biochem. J. **299**, 247 (1994).
458. GARCÍA-CARDENA, G., P. OH, J.W. LIU, J.E. SCHNITZER, and W.C. SESSA: Targeting of Nitric Oxide Synthase to Endothelial Cell Caveolae *via* Palmitoylation: Implications for Nitric Oxide Signaling. Proc. Natl. Acad. Sci. USA **93**, 6448 (1996).
459. SHAUL, P.W., E.J. SMART, L.J. ROBINSON, Z. GERMAN, I.S. YUHANNA, Y.S. YING, R.G.W. ANDERSON, and T. MICHEL: Acylation Targets Endothelial Nitric Oxide Synthase to Plasmalemmal Caveolae. J. Biol. Chem. **271**, 6518 (1996).
460. VENEMA, R.C., H.S. SAYEGH, J.F. ARNAL, and D.G. HARRISON: Role of the Enzyme Calmodulin-Binding Domain in Membrane Association and Phospholipid Inhibition of Endothelial Nitric Oxide Synthase. J. Biol. Chem. **270**, 14705 (1995).
461. LIU, J.W., and W.C. SESSA: Identification of Covalently Bound Amino-Terminal Myristic Acid in Endothelial Nitric Oxide Synthase. J. Biol. Chem. **269**, 11691 (1994).
462. BUSCONI, L., and T. MICHEL: Endothelial Nitric Oxide Synthase Membrane Targeting: Evidence Against Involvement of a Specific Myristate Receptor. J. Biol. Chem. **269**, 25016 (1994).
463. ROBINSON, L.J., and T. MICHEL: Mutagenesis of Palmitoylation Sites in Endothelial Nitric Oxide Synthase Identifies a Novel Motif for Dual Acylation and Subcellular Targeting. Proc. Natl. Acad. Sci. USA **92**, 11776 (1995).
464. OHASHI, Y., M. KATAYAMA, K. HIRATA, M. SUEMATSU, S. KAWASHIMA, and M. YOKOYAMA: Activation of Nitric Oxide Synthase from Cultured Aortic Endothelial Cells by Phospholipids. Biochem. Biophys. Res. Commun. **195**, 1314 (1993).
465. CALDERON, C., Z.H. HUANG, D.A. GAGE, E.M. SOTOMAYOR, and D.M. LOPEZ: Isolation of a Nitric Oxide Inhibitor from Mammary Tumor Cells and Its Characterization as Phosphatidyl Serine. J. Exp. Med. **180**, 945 (1994).
466. HIRATA, K., M. KATAYAMA, Y. OHASHI, R. KURODA, M. SUEMATSU, S. KAWASHIMA, and M. YOKOYAMA: Activation and Inhibition of Nitric Oxide Synthase from Cultured Bovine Aortic Endothelial Cells by Phospholipids and Arachidonic Acid. Annals of the New York Academy of Sciences **748**, 555 (1995).
467. ARAMAKI, Y., F. NITTA, R. MATSUNO, Y. MORIMURA, and S. TSUCHIYA: Inhibitory Effects of Negatively Charged Liposomes on Nitric Oxide Production from Macrophages Stimulated by LPS. Biochem. Biophys. Res. Commun. **220**, 1 (1996).
468. LIU, J.W., G. GARCÍA-CARDEÑA, and W.C. SESSA: Palmitoylation of Endothelial Nitric Oxide Synthase Is Necessary for Optimal Stimulated Release of Nitric Oxide: Implications for Caveolae Localization. Biochemistry **35**, 13277 (1996).
469. MICHEL, J.B., and T. MICHEL: The Role of Palmitoyl-Protein Thioesterase in the Palmitoylation of Endothelial Nitric Oxide Synthase. FEBS Letters **405**, 356 (1997).
470. LIU, J.W., T.E. HUGHES, and W.C. SESSA: The First 35 Amino Acids and Fatty Acylation Sites Determine the Molecular Targeting of Endothelial Nitric Oxide Synthase Into the Golgi Region of Cells: A Green Fluorescent Protein Study. Journal of Cell Biology **137**, 1525 (1997).
471. MICHEL, J.B., O. FERON, D. SACKS, and T. MICHEL: Reciprocal Regulation of Endothelial Nitric Oxide Synthase by Ca^{2+}-Calmodulin and Caveolin. J. Biol. Chem. **272**, 15583 (1997).
472. SZABO, C.: Physiological and Pathophysiological Roles of Nitric Oxide in the Central Nervous System. Brain Res. Bull. **41**, 131 (1996).

473. Dawson, T.M., V.L. Dawson, and S.H. Snyder: A Novel Neuronal Messenger Molecule in Brain: the Free Radical, Nitric Oxide. Annals of Neurology **32**, 297 (1992).

474. Hantraye, P., E. Brouillet, R. Ferrante, S. Palfi, R. Dolan, R.T. Matthews, and M.F. Beal: Inhibition of Neuronal Nitric Oxide Synthase Prevents MPTP-Induced Parkinsonism in Baboons. Nature Medicine **2**, 1017 (1996).

475. Huang, Z.H., P.L. Huang, N. Panahian, T. Dalkara, M.C. Fishman, and M.A. Moskowitz: Effects of Cerebral Ischemia in Mice Deficient in Neuronal Nitric Oxide Synthase. Science **265**, 1883 (1994).

476. Deliconstantinos, G., V. Villiotou, and J.C. Stavrides: Modulation of Particulate Nitric Oxide Synthase Activity and Peroxynitrite Synthesis in Cholesterol-Enriched Endothelial Cell Membranes. Biochem. Pharmac. **49**, 1589 (1995).

477. Wolff, D.J., G.A. Datto, and R.A. Samatovicz: The Dual Mode of Inhibition of Calmodulin-Dependent Nitric Oxide Synthase by Antifungal Imidazole Agents. J. Biol. Chem. **268**, 9430 (1993).

478. Wang, J.L., D.J. Stuehr, M. Ikeda-Saito, and D.L. Rousseau: Heme Coordination and Structure of the Catalytic Site in Nitric Oxide Synthase. J. Biol. Chem. **268**, 22255 (1993).

479. Degtyarenko, K.N., and A.I. Archakov: Molecular Evolution of P450 Superfamily and P450-Containing Monooxygenase Systems. FEBS Letters **332**, 1 (1993).

480. Sari, M.A., S. Booker, M. Jaouen, S. Vadon, J.L. Boucher, D. Pompon, and D. Mansuy: Expression in Yeast and Purification of Functional Macrophage Nitric Oxide Synthase: Evidence for Cysteine-194 as Iron Proximal Ligand. Biochemistry **35**, 7204 (1996).

481. Wang, J.L., D.J. Stuehr, and D.L. Rousseau: Tetrahydrobiopterin-Deficient Nitric Oxide Synthase Has a Modified Heme Environment and Forms a Cytochrome P-420 Analogue. Biochemistry **34**, 7080 (1995).

482. McMillan, K., and B.S.S. Masters: Optical Difference Spectrophotometry as a Probe of Rat Brain Nitric Oxide Synthase Heme-Substrate Interaction. Biochemistry **32**, 9875 (1993).

483. Sono, M., D.J. Stuehr, M. Ikeda-Saito, and J.H. Dawson: Identification of Nitric Oxide Synthase as a Thiolate-Ligated Heme Protein Using Magnetic Circular Dichroism Spectroscopy: Comparison with Cytochrome P-450$_{CAM}$ and Chloroperoxidase. J. Biol. Chem. **270**, 19943 (1995).

484. Wang, J.L., D.L. Rousseau, H.M. Abu-Soud, and D.J. Stuehr: Heme Coordination of NO in NO Synthase. Proc. Natl. Acad. Sci. USA **91**, 10512 (1994).

485. Salerno, J.C., C. Frey, K. McMillan, R.F. Williams, B.S.S. Masters, and O.W. Griffith: Characterization by Electron Paramagnetic Resonance of the Interactions of L-Arginine and L-Thiocitrulline with the Heme Cofactor Region of Nitric Oxide Synthase. J. Biol. Chem. **270**, 27423 (1995).

486. Salerno, J.C., K. McMillan, and B.S.S. Masters: Binding of Intermediate, Product, and Substrate Analogs to Neuronal Nitric Oxide Synthase: Ferriheme Is Sensitive to Ligand Specific Effects in the L-Arginine Binding Site. Biochemistry **35**, 11839 (1996).

487. Rodriguez-Crespo, I., N.C. Gerber, and P.R.O. Demontellano: Endothelial Nitric Oxide Synthase: Expression in *Escherichia Coli*, Spectroscopic Characterization, and Role of Tetrahydrobiopterin in Dimer Formation. J. Biol. Chem. **271**, 11462 (1996).

488. Chabin, R.M., E. McCauley, J.R. Calaycay, T.M. Kelly, K.L. Macnaul, G.C. Wolfe, N.I. Hutchinson, S. Madhusudanaraju, J.A. Schmidt, J.W. Kozarich, and K.K. Wong: Active Site Structure Analysis of Recombinant Human Inducible Nitric Oxide Synthase Using Imidazole. Biochemistry **35**, 9567 (1996).

489. ROMAN, L.J., E.A. SHETA, P. MARTASEK, S.S. GROSS, Q. LIU, and B.S.S. MASTERS: High-Level Expression of Functional Rat Neuronal Nitric Oxide Synthase in *Escherichia coli*. Proc. Natl. Acad. Sci. USA **92**, 8428 (1995).

490. PRESTA, A., U. SIDDHARTA, C.Q. WU, N. SENNEQUIER, L.X. HUANG, H.M. ABU-SOUD, S. ERZURUM, and D.J. STUEHR: Comparative Functioning of Dihydro- and Tetrahydropterins in Supporting Electron Transfer, Catalysis, and Subunit Dimerization in Inducible Nitric Oxide Synthase. Biochemistry **37**, 298 (1998).

491. MARTASEK, P., Q. LIU, J.W. LIU, L.J. ROMAN, S.S. GROSS, W.C. SESSA, and B.S.S. MASTERS: Characterization of Bovine Endothelial Nitric Oxide Synthase Expressed in *Escherichia Coli*. Biochem. Biophys. Res. Commun. **219**, 359 (1996).

492. FOSSETTA, J.D., X.D. NIU, C.A. LUNN, P.J. ZAVODNY, S.K. NARULA, and D. LUNDELL: Expression of Human Inducible Nitric Oxide Synthase in *Escherichia Coli*. FEBS Letters **379**, 135 (1996).

493. SENNEQUIER, N., and D.J. STUEHR: Analysis of Substrate-Induced Electronic, Catalytic, and Structural Changes in Inducible NO Synthase. Biochemistry **35**, 5883 (1996).

494. WOLFF, D.J., G.A. DATTO, R.A. SAMATOVICZ, and R.A. TEMPSICK: Calmodulin-Dependent Nitric Oxide Synthase – Mechanism of Inhibition by Imidazole and Phenylimidazoles. J. Biol. Chem. **268**, 9425 (1993).

495. WOLFF, D.J., and B.J. GRIBIN: The Inhibition of the Constitutive and Inducible Nitric Oxide Synthase Isoforms by Indazole Agents. Arch. Biochem. Biophys. **311**, 300 (1994).

496. WOLFF, D.J., A. LUBESKIE, and S. UMANSKY: The Inhibition of the Constitutive Bovine Endothelial Nitric Oxide Synthase by Imidazole and Indazole Agents. Arch. Biochem. Biophys. **314**, 360 (1994).

497. WOLFF, D.J., and B.J. GRIBIN: Interferon-γ-Inducible Murine Macrophage Nitric Oxide Synthase – Studies on the Mechanism of Inhibition by Imidazole Agents. Arch. Biochem. Biophys. **311**, 293 (1994).

498. MAYER, B., P. KLATT, E.R. WERNER, and K. SCHMIDT: Identification of Imidazole as L-Arginine-Competitive Inhibitor of Porcine Brain Nitric Oxide Synthase. FEBS Letters **350**, 199 (1994).

499. COUNTS GERBER, N., P. R. ORTIZ DE MONTELLANO: Neuronal Nitric Oxide Synthase: Expression in *Escherichia coli*, Irreversible Inhibition by Phenyldiazene, and Active Site Topology. J. Biol. Chem. **270**, 17791 (1995).

500. COUNTS GERBER, N., I. RODRIGUEZ-CRESPO, C.R. NISHIDA, and P.R. ORTIZ DE MONTELLANO: Active Site Topologies and Cofactor-Mediated Conformational Changes of Nitric Oxide Synthases. J. Biol. Chem. **272**, 6285 (1997).

501. KLATT, P., M. SCHMID, E. LEOPOLD, K. SCHMIDT, E.R. WERNER, and B. MAYER: The Pteridine Binding Site of Brain Nitric Oxide Synthase: Tetrahydrobiopterin Binding Kinetics, Specificity, and Allosteric Interaction with the Substrate Domain. J. Biol. Chem. **269**, 13861 (1994).

502. BERKA, V., P.-F. CHEN, and A.-L. TSAI: Spatial Relationship between L-Arginine and Heme Binding Sites of Endothelial Nitric Oxide Synthase. J. Biol. Chem. **271**, 33293 (1996).

503. MICHEL, A.D., R.K. PHUL, T.L. STEWART, and P.P.A. HUMPHREY: Characterization of the Binding of [H-3] L-N^G-Nitroarginine in Rat Brain. Br. J. Pharmac. **109**, 287 (1993).

504. KLATT, P., K. SCHMIDT, F. BRUNNER, and B. MAYER: Inhibitors of Brain Nitric Oxide Synthase – Binding-Kinetics, Metabolism, and Enzyme Inactivation. J. Biol. Chem. **269**, 1674 (1994).

505. GARVEY, E.P., J.A. OPLINGER, G.J. TANOURY, P.A. SHERMAN, M. FOWLER, S. MARSHALL, M.F. HARMON, J.E. PAITH, and E.S. FURFINE: Potent and Selective Inhibition of Human

Nitric Oxide Synthases – Inhibition by Non-Amino Acid Isothioureas. J. Biol. Chem. **269**, 26669 (1994).

506. Furfine, E.S., M.F. Harmon, J.E. Paith, R.G. Knowles, M. Salter, R.J. Kiff, C. Duffy, R. Hazelwood, J.A. Oplinger, and E.P. Garvey: Potent and Selective Inhibition of Human Nitric Oxide Synthases – Selective Inhibition of Neuronal Nitric Oxide Synthase by *S*-methyl-L-Thiocitrulline and *S*-Ethyl-L-Thiocitrulline. J. Biol. Chem. **269**, 26677 (1994).

507. Fukuto, J.M., K.S. Wood, R.E. Byrns, and L.J. Ignarro: N^G-Amino-L-arginine: A New Potent Antagonist of L-Arginine-Mediated Endothelium-Dependent Relaxation. Biochem. Biophys. Res. Commun. **168**, 458 (1990).

508. McCall, T.B., M. Feelisch, R.M.J. Palmer, and S. Moncada: Identification of *N*-Iminoethyl-L-Ornithine as an Irreversible Inhibitor of Nitric Oxide Synthase in Phagocytic Cells. Br. J. Pharmac. **102**, 234 (1991).

509. Frey, C., K. Narayanan, K. McMillan, L. Spack, S.S. Gross, B.S. Masters, and O.W. Griffith: L-Thiocitrulline – a Stereospecific, Heme-Binding Inhibitor of Nitric Oxide Synthases. J. Biol. Chem. **269**, 26083 (1994).

510. Nishimura, J.S., P. Martasek, K. McMillan, J.C. Salerno, Q. Liu, S.S. Gross, and B.S.S. Masters: Modular Structure of Neuronal Nitric Oxide Synthase – Localization of the Arginine Binding Site and Modulation by Pterin. Biochem. Biophys. Res. Commun. **210**, 288 (1995).

511. Elkasmi, A., D. Lexa, P. Maillard, M. Momenteau, and J.M. Saveant: Reversible Two-Electron One-Proton Systems in the Ring-Centered Oxidation of Metalloporphyrins Bearing Secondary Amide Linked Superstructures. J. Am. Chem. Soc. **113**, 1586 (1991).

512. Chang, C.K., G. Aviles, and N.Z. Bag: Verdoheme-Like Oxaporphyrin Formation by Oxygenation of a Co(II) Porphyrinyl Naphthoic Acid: A New Model of Heme Degradation. J. Am. Chem. Soc. **116**, 12127 (1994).

513. Kanyo, Z.F., L.R. Scolnick, D.E. Ash, and D.W. Christianson: Structure of a Unique Binuclear Manganese Cluster in Arginase. Nature **383**, 554 (1996).

514. Ishikawa, T., T. Misonou, M. Ikeno, K. Sato, and T. Sakamaki: N^ω-Hydroxyagmatine – a Novel Substance Causing Endothelium-Dependent Vasorelaxation. Biochem. Biophys. Res. Commun. **214**, 145 (1995).

515. Lee, C.M., L.J. Robinson, and T. Michel: Oligomerization of Endothelial Nitric Oxide Synthase: Evidence for a Dominant Negative Effect of Truncation Mutants. J. Biol. Chem. **270**, 27403 (1995).

516. Siddhanta, U., C.Q. Wu, H.M. Abu-Soud, J.L. Zhang, D.K. Ghosh, and D.J. Stuehr: Heme Iron Reduction and Catalysis by a Nitric Oxide Synthase Heterodimer Containing One Reductase and Two Oxygenase Domains. J. Biol. Chem. **271**, 7309 (1996).

517. Gorren, A.C.F., A. Schrammel, K. Schmidt, and B. Mayer: Thiols and Neuronal Nitric Oxide Synthase: Complex Formation, Competitive Inhibition, and Enzyme Stabilization. Biochemistry **36**, 4360 (1997).

518. List, B.M., P. Klatt, E.R. Werner, K. Schmidt, and B. Mayer: Overexpression of Neuronal Nitric Oxide Synthase in Insect Cells Reveals Requirement of Heme for Tetrahydrobiopterin Binding. Biochem. J. **315**, 57 (1996).

519. List, B.M., B. Klosch, C. Volker, A.C.F. Gorren, W.C. Sessa, E.R. Werner, W.R. Kukovetz, K. Schmidt, and B. Mayer: Characterization of Bovine Endothelial Nitric Oxide Synthase as a Homodimer with Down-Regulated Uncoupled NADPH Oxidase Activity: Tetrahydrobiopterin Binding Kinetics and Role of Haem in Dimerization. Biochem. J. **323**, 159 (1997).

520. GORREN, A.C.F., B.M. LIST, A. SCHRAMMEL, E. PITTERS, B. HEMMENS, E.R. WERNER, K. SCHMIDT, and B. MAYER: Tetrahydrobiopterin-Free Neuronal Nitric Oxide Synthase: Evidence for Two Identical Highly Anticooperative Pteridine Binding Sites. Biochemistry **35**, 16735 (1996).

521. FURFINE, E.S., M.F. HARMON, J.E. PAITH, and E.P. GARVEY: Selective Inhibition of Constitutive Nitric Oxide Synthase by L-N^G-Nitroarginine. Biochemistry **32**, 8512 (1993).

522. HARTENECK, C., P. KLATT, K. SCHMIDT, and B. MAYER: Expression of Rat Brain Nitric Oxide Synthase in Baculovirus-Infected Insect Cells and Characterization of the Purified Enzyme. Biochem. J. **304**, 683 (1994).

523. KLATT, P., S. PFEIFFER, B.M. LIST, D. LEHNER, O. GLATTER, H.P. BACHINGER, E.R. WERNER, K. SCHMIDT, and B. MAYER: Characterization of Heme-Deficient Neuronal Nitric Oxide Synthase Reveals a Role for Heme in Subunit Dimerization and Binding of the Amino-Acid Substrate and Tetrahydrobiopterin. J. Biol. Chem. **271**, 7336 (1996).

524. DJORDJEVIC, S., D.L. ROBERTS, M. WANG, T. SHEA, M.G.W. CAMITTA, B.S.S. MASTERS, and J.J.P. KIM: Crystallization and Preliminary X-Ray Studies of NADPH-Cytochrome P450 Reductase. Proc. Natl. Acad. Sci. USA **92**, 3214 (1995).

525. GROVES, J.T., and C.Y. WANG: Models for Nitric Oxide Synthase. Abstracts of Papers of the American Chemical Society **213**, 321 (1997).

526. KNOWLES, R.G., M. PALACIOS, R.M.J. PALMER, and S. MONCADA: Kinetic Characteristics of Nitric Oxide Synthase from Rat-Brain. Biochem. J. **269**, 207 (1990).

527. DWYER, M.A., D.S. BREDT, and S.H. SNYDER: Nitric Oxide Synthase – Irreversible Inhibition by L-N^G-Nitroarginine in Brain *In Vitro* and *In Vivo*. Biochem. Biophys. Res. Commun. **176**, 1136 (1991).

528. WERNER-FELMAYER, G., G. GOLDERER, E.R. WERNER, P. GROBNER, and H. WACHTER: Pteridine Biosynthesis and Nitric Oxide Synthase in *Physarum polycephalum*. Biochem. J. **304**, 105 (1994).

529. POLLOCK, J.S., E.R. WERNER, J.A. MITCHELL, and U. FÖRSTERMANN: Particulate Endothelial Nitric Oxide Synthase: Requirement and Content of Tetrahydrobiopterin, FAD, and FMN. Endothelium **1**, 147 (1993).

530. GROSS, S.S., E.A. JAFFE, R. LEVI, and R.G. KILBOURN: Cytokine-Activated Endothelial Cells Express an Isotype of Nitric Oxide Synthase Which Is Tetrahydrobiopterin-Dependent, Calmodulin-Independent and Inhibited by Arginine Analogs with a Rank Order of Potency Characteristic of Activated Macrophages. Biochem. Biophys. Res. Commun. **178**, 823 (1991).

531. WERNER-FELMAYER, G., E.R. WERNER, D. FUCHS, A. HAUSEN, G. REIBNEGGER, K. SCHMIDT, G. WEISS, and H. WACHTER: Pteridine Biosynthesis in Human Endothelial Cells – Impact on Nitric Oxide-Mediated Formation of Cyclic-GMP. J. Biol. Chem. **268**, 1842 (1993).

532. WEISS, G., B. GOOSSEN, W. DOPPLER, D. FUCHS, K. PANTOPOULOS, G. WERNER-FELMAYER, H. WACHTER, and M.W. HENTZE: Translational Regulation *via* Iron-Responsive Elements by the Nitric Oxide NO Synthase Pathway. EMBO J. **12**, 3651 (1993).

533. MELLOUK, S., S.L. HOFFMAN, Z.Z. LIU, P. DELAVEGA, T.R. BILLIAR, and A.K. NUSSLER: Nitric Oxide-Mediated Antiplasmodial Activity in Human and Murine Hepatocytes Induced by Gamma-Interferon and the Parasite Itself – Enhancement by Exogenous Tetrahydrobiopterin. Infection and Immunity **62**, 4043 (1994).

534. WERNER, E.R., G. WERNER-FELMAYER, and H. WACHTER: Tetrahydrobiopterin and Cytokines. Proc. Soc. Exp. Biol. Med. **203**, 1 (1993).

535. Werner, E.R., G. Werner-Felmayer, H. Wachter, and B. Mayer: Biosynthesis of Nitric Oxide: Dependence on Pteridine Metabolism. Reviews of Physiology Biochemistry and Pharmacology **127**, 97 (1996).

536. Giovanelli, J., K.L. Campos, and S. Kaufman: Tetrahydrobiopterin, a Cofactor for Rat Cerebellar Nitric Oxide Synthase, does not Function as a Reactant in the Oxygenation of Arginine. Proc. Natl. Acad. Sci. USA **88**, 7091 (1991).

537. Werner, E.R., E. Pitters, K. Schmidt, H. Wachter, G. Werner-Felmayer, and B. Mayer: Identification of the 4-Amino Analogue of Tetrahydrobiopterin as a Dihydropteridine Reductase Inhibitor and a Potent Pteridine Antagonist of Rat Neuronal Nitric Oxide Synthase. Biochem. J. **320**, 193 (1996).

538. Witteveen, C.F.B., J. Giovanelli, and S. Kaufman: Reduction of Quinonoid Dihydrobiopterin to Tetrahydrobiopterin by Nitric Oxide Synthase. J. Biol. Chem. **271**, 4143 (1996).

539. Schenk, H., H. Hofmann, L. Fröhlich, M. Meineke, M. Weeger, R. Weinberg, D. Marecak, G. Prestwich, V. Groehn, W. Pfleiderer, and H.H.H.W. Schmidt: The Nitric Oxide Synthase: Pterin Binding Site and Associated Reductase Activity. Naunyn-Schmiedeberg's Arch. Pharmacol. **355**, 148 (1997).

540. Mayer, B., B. Heinzel, P. Klatt, M. John, K. Schmidt, and E. Bohme: Nitric Oxide Synthase-Catalyzed Activation of Oxygen and Reduction of Cytochromes – Reaction Mechanisms and Possible Physiological Implications. J. Cardiovasc. Pharmac. **20**, S 54 (1992).

541. Griscavage, J.M., J.M. Fukuto, Y. Komori, and L.J. Ignarro: Nitric Oxide Inhibits Neuronal Nitric Oxide Synthase by Interacting with the Heme Prosthetic Group: Role of Tetrahydrobiopterin in Modulating the Inhibitory Action of Nitric Oxide. J. Biol. Chem. **269**, 21644 (1994).

542. Hyun, J., Y. Komori, G. Chaudhuri, L.J. Ignarro, and J.M. Fukuto: The Protective Effect of Tetrahydrobiopterin on the Nitric Oxide-Mediated Inhibition of Purified Nitric Oxide Synthase. Biochem. Biophys. Res. Commun. **206**, 380 (1995).

543. Mayer, B., P. Klatt, E.R. Werner, and K. Schmidt: Kinetics and Mechanism of Tetrahydrobiopterin-Induced Oxidation of Nitric Oxide. J. Biol. Chem. **270**, 655 (1995).

544. Uvarov, V.Y., and A.A. Lyashenko: The Identification of the Pterin-Binding Domain in the Nitric Oxide Synthases Sequence. Biochem. Biophys. Res. Commun. **206**, 736 (1995).

545. Davies, J.F., T.J. Delcamp, N.J. Prendergast, V.A. Ashford, J.H. Freisheim, and J. Kraut: Crystal Structures of Recombinant Human Dihydrofolate Reductase Complexed with Folate and 5-Deazafolate. Biochemistry **29**, 9467 (1990).

546. Bystroff, C., S.J. Oatley, and J. Kraut: Crystal Structures of *Escherichia coli* Dihydrofolate Reductase: the NADP+ Holoenzyme and the Folate-NADP+ Ternary Complex: Substrate Binding and a Model for the Transition State. Biochemistry **29**, 3263 (1990).

547. Marletta, M.A.: Approaches Toward Selective Inhibition of Nitric Oxide Synthase. J. Med. Chem. **37**, 1899 (1994).

548. Mayer, B., P. Klatt, E.R. Werner, and K. Schmidt: Molecular Mechanisms of Inhibition of Porcine Brain Nitric Oxide Synthase by the Antinociceptive Drug 7-Nitro-Indazole. Neuropharmacology **33**, 1253 (1994).

549. Frey, A., H. Schenk, L. Fröhlich, A. Reif-Russanoff, W. Pfleiderer, and H.H.H.W. Schmidt: Tetrahydrobiopterin-Analoga (Anti-Pterins) as Inhibitors of Human NO Synthases. Naunyn-Schmiedeberg's Arch. Pharmacol. **355**, 147 (1997).

550. GHOSH, D.K., H.M. ABU-SOUD, and D.J. STUEHR: Domains of Macrophage NO Synthase Have Divergent Roles in Forming and Stabilizing the Active Dimeric Enzyme. Biochemistry **35**, 1444 (1996).

551. ABU-SOUD, H.M., M. LOFTUS, and D.J. STUEHR: Subunit Dissociation and Unfolding of Macrophage NO Synthase: Relationship Between Enzyme Structure, Prosthetic Group Binding, and Catalytic Function. Biochemistry **34**, 11167 (1995).

552. TÓTH, M., Z. KUKOR, and M. SAHIN-TÓTH: Activation and Dimerization of Type III Nitric Oxide Synthase by Submicromolar Concentrations of Tetrahydrobiopterin in Microsomal Preparations from Human Primordial Placenta. Placenta **18**, 189 (1997).

553. HELLERMANN, G.R., and L.P. SOLOMONSON: Calmodulin Promotes Dimerization of the Oxygenase Domain of Human Endothelial Nitric Oxide Synthase. J. Biol. Chem. **272**, 12030 (1997).

554. ASSREUY, J., F.Q. CUNHA, F.Y. LIEW, and S. MONCADA: Feedback Inhibition of Nitric Oxide Synthase Activity by Nitric Oxide. Br. J. Pharmac. **108**, 833 (1993).

555. RENGASAMY, A., and R.A. JOHNS: Regulation of Nitric Oxide Synthase by Nitric Oxide. Molec. Pharmac. **44**, 124 (1993).

556. ROGERS, N.E., and L.J. IGNARRO: Constitutive Nitric Oxide Synthase from Cerebellum Is Reversibly Inhibited by Nitric Oxide Formed from L-Arginine. Biochem. Biophys. Res. Commun. **189**, 242 (1992).

557. BUGA, G.M., J.M. GRISCAVAGE, N.E. ROGERS, and L.J. IGNARRO: Negative Feedback Regulation of Endothelial Cell Function by Nitric Oxide. Circ. Res. **73**, 808 (1993).

558. GRISCAVAGE, J.M., N.E. ROGERS, M.P. SHERMAN, and L.J. IGNARRO: Inducible Nitric Oxide Synthase from a Rat Alveolar Macrophage Cell Line Is Inhibited by Nitric Oxide. J. Immunol. **151**, 6329 (1993).

559. RAVICHANDRAN, L.V., R.A. JOHNS, and A. RENGASAMY: Direct and Reversible Inhibition of Endothelial Nitric Oxide Synthase by Nitric Oxide. Am. J. Physiol. **37**, H2216 (1995).

560. HURSHMAN, A.R., and M.A. MARLETTA: Nitric Oxide Complexes of Inducible Nitric Oxide Synthase – Spectral Characterization and Effect on Catalytic Activity. Biochemistry **34**, 5627 (1995).

561. ABU-SOUD, H.M., J.L. WANG, D.L. ROUSSEAU, J.M. FUKUTO, L.J. IGNARRO, and D.J. STUEHR: Neuronal Nitric Oxide Synthase Self-Inactivates by Forming a Ferrous-Nitrosyl Complex during Aerobic Catalysis. J. Biol. Chem. **270**, 22997 (1995).

562. ABU-SOUD, H.M., D.L. ROUSSEAU, and D.J. STUEHR: Nitric Oxide Binding to the Heme of Neuronal Nitric Oxide Synthase Links Its Activity to Changes in Oxygen Tension. J. Biol. Chem. **271**, 32515 (1996).

563. GRUETTER, C.A., B.K. BARRY, D.B. MCNAMARA, D.Y. GRUETTER, P.J. KADOWITZ, and L.J. IGNARRO: Relaxation of Bovine Coronary Artery and Activation of Coronary Arterial Guanylate Cyclase by Nitric Oxide, Nitroprusside and a Carcinogenic Nitrosamine. J. Cyclic Nucleotide Res. **5**, 211 (1979).

564. YONETANI, T., H. YAMAMOTO, J.E. ERMAN, J.S. LEIGH, and G.H. REED: Electromagnetic Properties of the Hemoproteins. V. Optical and Electron Paramagnetic Resonance Characteristics of Nitric Oxide Derivatives of Metalloporphyrin-Apohemoprotein Complexes. J. Biol. Chem. **247**, 2447 (1972).

565. O'KEEFE, D.H., R.E. EBEL, and J.A. PETERSON: Studies of the Oxygen Binding Site of Cytochrome P-450: Nitric Oxide as a Spin Label Probe. J. Biol. Chem. **253**, 3509 (1978).

566. SHARMA, V.S., T.G. TRAYLOR, R. GARDINER, and H. MIZUKAMI: Reaction of Nitric Oxide with Heme Proteins and Model Compounds of Hemoglobin. Biochemistry **26**, 3837 (1987).

567. ARNOLD, W.P., C.K. MITTAL, S. KATSUKI, and F. MURAD: Nitric Oxide Activates Guanylate Cyclase and Increases Guanosine $3':5'$-Cyclic Monophosphate Levels in Various Tissue Preparations. Proc. Natl. Acad. Sci. USA **74**, 3203 (1977).

568. CRAVEN, P.A., and F.R. DERUBERTIS: Restoration of the Responsiveness of Purified Guanylate Cyclase to Nitrosoguanidine, Nitric Oxide, and Related Activators by Heme and Hemeproteins: Evidence for Involvement of the Paramagnetic Nitrosyl-Heme Complex in Enzyme Activation. J. Biol. Chem. **253**, 8433 (1978).

569. PATEL, J.M., and E.R. BLOCK: Sulfhydryl-Disulfide Modulation and the Role of Disulfide Oxidoreductases in Regulation of the Catalytic Activity of Nitric Oxide Synthase in Pulmonary Artery Endothelial Cells. Am. J. Respiratory Cell Molec. Biol. **13**, 352 (1995).

570. PATEL, J.M., J.L. ZHANG, and E.R. BLOCK: Nitric Oxide–Induced Inhibition of Lung Endothelial Cell Nitric Oxide Synthase Via Interaction with Allosteric Thiols – Role of Thioredoxin in Regulation of Catalytic Activity. Am. J. Respiratory Cell Molec. Biol. **15**, 410 (1996).

571. PARK, S.K., H.L. LIN, and S. MURPHY: Nitric Oxide Regulates Nitric Oxide Synthase-2 Gene Expression by Inhibiting NF-Kappa B Binding to DNA. Biochem. J. **322**, 609 (1997).

572. FÖRSTERMANN, U., and H. KLEINERT: Nitric Oxide Synthase: Expression and Expressional Control of the 3 Isoforms. Naunyn-Schmiedeberg's Arch. Pharmacol. **352**, 351 (1995).

573. LOWENSTEIN, C.J., E.W. ALLEY, P. RAVAL, A.M. SNOWMAN, S.H. SNYDER, S.W. RUSSELL, and W.J. MURPHY: Macrophage Nitric Oxide Synthase Gene: Two Upstream Regions Mediate Induction by Interferon Gamma and Lipopolysaccharide. Proc. Natl. Acad. Sci. USA **90**, 9730 (1993).

574. CULCASI, M., M. LAFONCAZAL, S. PIETRI, and J. BOCKAERT: Glutamate Receptors Induce a Burst of Superoxide Via Activation of Nitric Oxide Synthase in Arginine-Depleted Neurons. J. Biol. Chem. **269**, 12589 (1994).

575. XIA, Y., and J.L. ZWEIER: Superoxide and Peroxynitrite Generation from Inducible Nitric Oxide Synthase in Macrophages. Proc. Natl. Acad. Sci. USA **94**, 6954 (1997).

576. PASQUET, J.P.E.E., M.H. ZOU, and V. ULLRICH: Peroxynitrite Inhibition of Nitric Oxide Synthases. Biochimie **78**, 785 (1996).

577. HUHMER, A.F.R., C.R. NISHIDA, P.R.O. DEMONTELLANO, and C. SCHONEICH: Inactivation of the Inducible Nitric Oxide Synthase by Peroxynitrite. Chem. Res. Toxicol. **10**, 618 (1997).

578. KOMORI, Y., J.J. HYUN, K. CHIANG, and J.M. FUKUTO: The Role of Thiols in the Apparent Activation of Rat-Brain Nitric Oxide Synthase (NOS). J. Biochem. **117**, 923 (1995).

579. HOFMANN, H., and H.H.H.W. SCHMIDT: Thiol Dependence of Nitric Oxide Synthase. Biochemistry **34**, 13443 (1995).

580. STEWART, D.J.: Endothelial Dysfunctions in Pulmonary Vascular Diseases. Arzemittelforschung **44**, 451 (1994).

581. FROSTELL, C.G., H. BLOMQVIST, G. HEDENSTIERNA, J. LUNDBERG, and W.M. ZAPOL: Inhaled Nitric Oxide Selectively Reverses Human Hypoxic Pulmonary Vasoconstriction without Causing Systemic Vasodilation. Anesthesiology **78**, 427 (1993).

582. FRATACCI, M.D., C.G. FROSTELL, T.Y. CHEN, J.C. WAIN, D.R. ROBINSON, and W.M. ZAPOL: Inhaled Nitric Oxide – A Selective Pulmonary Vasodilator of Heparin Protamine Vasoconstriction in Sheep. Anesthesiology **75**, 990 (1991).

583. Ogura, H., W.G. Cioffi, P.J. Offner, B.S. Jordan, A.A. Johnson, and B.A. Pruitt: Effect of Inhaled Nitric Oxide on Pulmonary Function after Sepsis in a Swine Model. Surgery **116**, 313 (1994).

584. Romand, J.A., M.R. Pinsky, L. Firestone, H.A. Zar, and J.R. Lancaster: Effect of Inhaled Nitric Oxide on Pulmonary Hemodynamics after Acute Lung Injury in Dogs. J. Appl. Physiol. **76**, 1356 (1994).

585. Tonz, M., L.K. Von Segesser, J. Schilling, T.F. Luscher, G. Noll, B. Leskosek, and M.I. Turina: Treatment of Acute Pulmonary Hypertension with Inhaled Nitric Oxide. Ann. Thorac. Surg. **58**, 1031 (1994).

586. Roberts, J.D., D.M. Polaner, P. Lang, and W.M. Zapol: Inhaled Nitric Oxide in Persistent Pulmonary Hypertension of the Newborn. Lancet **340**, 818 (1992).

587. Kinsella, J.P., S.R. Neish, E. Shaffer, and S.H. Abman: Low Dose Inhalational Nitric Oxide in Persistent Pulmonary Hypertension of the Newborn. Lancet **340**, 819 (1992).

588. Finer, N.N., P.C. Etches, B. Kamstra, A.J. Tierney, A. Peliowski, and C.A. Ryan: Inhaled Nitric Oxide in Infants Referred for Extracorporeal Membrane Oxygenation – Dose Response. J. Pediatr. **124**, 302 (1994).

589. Girard, C., J.J. Lehot, J.C. Pannetier, S. Filley, P. Ffrench, and S. Estanove: Inhaled Nitric Oxide after Mitral Valve Replacement in Patients with Chronic Pulmonary Artery Hypertension. Anesthesiology **77**, 880 (1992).

590. Mill R, O.I., D.S. Celermajer, J.E. Deanfield, and D.J. Macrae: Very Low Dose Inhaled Nitric Oxide – A Selective Pulmonary Vasodilator after Operations for Congenital Heart Disease. J. Thorac. Cardiovasc. Surg. **108**, 487 (1994).

591. Packer, M., B. Greenberg, B. Massie, and H. Dash: Deleterious Effects of Hydralazine in Patients with Pulmonary Hypertension. New Eng. J. Med. **306**, 1326 (1982).

592. Wessel, D.L., I. Adatia, T.M. Giglia, J.E. Thompson, and T.J. Kulik: Use of Inhaled Nitric Oxide and Acetylcholine in the Evaluation of Pulmonary Hypertension and Endothelial Function after Cardiopulmonary Bypass. Circulation **88**, I- 2128 (1993).

593. Delgado, R., A. Rojas, L.A. Glaria, M. Torres, F. Duarte, R. Shill, M. Nafeh, E. Santin, N. Gonzalez, and M. Palacios: Ca^{2+} Independent Nitric Oxide Synthase Activity in Human Lung after Cardiopulmonary Bypass. Thorax **50**, 403 (1995).

594. Davis, R.D., E.P. Trulock, J. Manley, M.K. Pasque, S. Sundaresan, J.D. Cooper, A. Patterson, B.P. Griffith, T.M. Egan, and T.R. Todd: Differences in Early Results after Single Lung Transplantation. Ann. Thorac. Surg. **58**, 1327 (1994).

595. Kirk, A.J.B., I.W. Colquhoun, and J.H. Dark: Lung Preservation – A Review of Current Practice and Future Directions. Ann. Thorac. Surg. **56**, 990 (1993).

596. Date, H., A.N. Triantafillou, E.P. Trulock, M.S. Pohl, J.D. Cooper, and G.A. Patterson: Inhaled Nitric Oxide Reduces Human Lung Allograft Dysfunction. J. Thorac. Cardiovasc. Surg. **111**, 913 (1996).

597. Shiraishi, T., S.R. Demeester, N.K. Worrall, J.H. Ritter, T.P. Misko, T.B. Ferguson, J.D. Cooper, and G.A. Patterson: Inhibition of Inducible Nitric Oxide Synthase Ameliorates Rat Lung Allograft Rejection. J. Thorac. Cardiovasc. Surg. **110**, 1449 (1995).

598. Worrall, N.K., T.P. Misko, P.M. Sullivan, J.J. Hui, and T.B. Ferguson: Inhibition of Inducible Nitric Oxide Synthase Attenuates Established Acute Cardiac Allograft Rejection. Ann. Thorac. Surg. **62**, 378 (1996).

599. Lewis, N.P., P.S. Tsao, P.R. Rickenbacher, C. Xue, R.A. Johns, G.A. Haywood, H. von der Leyen, P.T. Trindade, J.P. Cooke, S.A. Hunt, M.E. Billingham, H.A. Valantine, and M.B. Fowler: Induction of Nitric Oxide Synthase in the Human Cardiac Allograft is Associated with Contractile Dysfunction of the Left Ventricle. Circulation **93**, 720 (1996).

600. Barnes, P.J., and F.Y. Liew: Nitric Oxide and Asthmatic Inflammation. Immunol. Today **16**, 128 (1995).

601. Kharitonov, S.A., D. Yates, R.A. Robbins, R. Logan Sinclair, E.A. Shinebourne, and P.J. Barnes: Increased Nitric Oxide in Exhaled Air of Asthmatic Patients. Lancet **343**, 133 (1994).

602. Kharitonov, S.A., K.K. Chung, D. Evans, B.J. O'connor, and P.J. Barnes: Increased Exhaled Nitric Oxide in Asthma is Mainly Derived from the Lower Respiratory Tract. Am. J. Respir. Crit. Care Med. **153**, 1773 (1996).

603. Yates, D.H., S.A. Kharitonov, R.A. Robbins, P.S. Thomas, and P.J. Barnes: Effect of a Nitric Oxide Synthase Inhibitor and a Glucocorticosteroid on Exhaled Nitric Oxide. Am. J. Resp. Crit. Care Med. **152**, 892 (1995).

604. Yates, D.H., S.A. Kharitonov, P.S. Thomas, and P.J. Barnes: Endogenous Nitric-Oxide is Decreased in Asthmatic Patients by an Inhibitor of Inducible Nitric Oxide Synthase. Am. J. Respir. Crit. Care Med. **154**, 247 (1996).

605. Hogman, M., C.G. Frostell, H. Hedenstrom, and G. Hedenstierna: Inhalation of Nitric Oxide Modulates Adult Human Bronchial Tone. Am. Rev. Respir. Dis. **148**, 1474 (1993).

606. Barnes, P.J.: NO or No NO in Asthma? Thorax **51**, 218 (1996).

607. Blanch, P.B., J.C. Koens, and A.J. Layon: A New Device that Allows Synchronous Intermittent Inspiratory Chest Tube Occlusion with any Mechanical Ventilator. Chest **97**, 1426 (1990).

608. Semigran, M.J., B.A. Cockrill, R. Kacmarek, B.T. Thompson, W.M. Zapol, G.W. Dec, and M.A. Fifer: Hemodynamic Effects of Inhaled Nitric Oxide in Heart Failure. J. Am. Coll. Cardiol. **24**, 982 (1994).

609. Bocchi, E.A., F. Bacal, J.O.C. Auler, M.J.D. Carmone, G. Bellotti, and F. Pileggi: Inhaled Nitric Oxide Leading to Pulmonary Edema in Stable Severe Heart Failure. Am. J. Cardiol. **74**, 70 (1994).

610. Morrow, P.E.: Toxicological Data on NOX – an Overview. J. Toxicol. Environ. Health **13**, 205 (1984).

611. Stewart, T.E., F. Valenza, S.P. Ribeiro, A.D. Wener, G. Volgyesi, J.B.M. Mullen, and A.S. Slutsky: Increased Nitric Oxide in Exhaled Gas as an Early Marker of Lung Inflammation in a Model of Sepsis. Am. J. Resp. Crit. Care Med. **151**, 713 (1995).

612. McHugh, L.G., J.A. Milberg, M.E. Whitcomb, R.B. Schoene, R.J. Maunder, and L.D. Hudson: Recovery of Function in Survivors of the Acute Respiratory Distress Syndrome. Am. J. Resp. Crit. Care Med. **150**, 90 (1994).

613. Bigatello, L.M., W.E. Hurford, R.M. Kacmarek, J.D. Roberts, and W.M. Zapol: Prolonged Inhalation of Low Concentrations of Nitric Oxide in Patients with Severe Adult Respiratory Distress Syndrome Effects on Pulmonary Hemodynamics and Oxygenation. Anesthesiology **80**, 761 (1994).

614. Pison, U., F.A. Lopez, C.F. Heidelmeyer, R. Rossaint, and K.J. Falke: Inhaled Nitric Oxide Reverses Hypoxic Pulmonary Vasoconstriction without Impairing Gas Exchange. J. Appl. Physiol. **74**, 1287 (1993).

615. Pepke-Zaba, J., T.W. Higenbottam, A.T. Dinhxuan, D. Stone, and J. Wallwork: Inhaled Nitric Oxide as a Cause of Selective Pulmonary Vasodilatation in Pulmonary Hypertension. Lancet **338**, 1173 (1991).

616. Fierobe, L., F. Brunet, J.F. Dhainaut, M. Monchi, M. Belghith, J.P. Mira, J. Dallavasantucci, and A.T. Dinhxuan: Effect of Inhaled Nitric Oxide on Right Ventricular Function in Adult Respiratory Distress Syndrome. Am. J. Resp. Crit. Care Med. **151**, 1414 (1995).

617. ROSSAINT, R., K.J. FALKE, F. LOPEZ, K. SLAMA, U. PISON, and W.M. ZAPOL: Inhaled Nitric Oxide for the Adult Respiratory Distress Syndrome. New Eng. J. Med. **328**, 399 (1993).

618. LEEMAN, M., V.Z. DEBEYL, E. GILBERT, C. MELOT, and R. NAEIJE: Is Nitric Oxide Released in Oleic Acid Lung Injury? J. Appl. Physiol. **74**, 650 (1993).

619. KAVANAGH, B.P., A. MOUCHAWAR, J. GOLDSMITH, and R.G. PEARL: Effects of Inhaled NO and Inhibition of Endogenous NO Synthesis in Oxidant Induced Acute Lung Injury. J. Appl. Physiol. **76**, 1324 (1994).

620. ROSSAINT, R., H. GERLACH, H. SCHMIDTRUHNKE, D. PAPPERT, K. LEWANDOWSKI, W. STEUDEL, and K. FALKE: Efficacy of Inhaled Nitric Oxide in Patients with Severe ARDS. Chest **107**, 1107 (1995).

621. SUCHYTA, M.R., T.P. CLEMMER, C.G. ELLIOTT, J.F. ORME, and L.K. WEAVER: The Adult Respiratory Distress Syndrome – A Report of Survival and Modifying Factors. Chest **101**, 1074 (1992).

622. BONE, R.C.: The Pathogenesis of Sepsis. Ann. Int. Med. **115**, 457 (1991).

623. FLEMING, I., G.A. GRAY, G. JULON-SHAEFFER, J.R. PARRATT, and J-C. STOCLET: Incusation with Endotoxin Activates the L-Arginine Pathway in Vascular Tisssue. Biochem. Biophys. Res, Commun. **171**, 562 (1990).

624. WESTENBERGER, U., S. THANNER, H.H. RUF, K. GERSONDE, G. SUTTER, and O. TRENTZ: Formation of Free Radicals and Nitric Oxide Derivative of Hemoglobin in Rats During Shock Syndrome. Free Rad. Res. Comm. **11**, 167 (1990).

625. REILING, N., A.J. ULMER, M. DUCHROW, M. ERNST, H.D. FLAD, and S. HAUSCHILDT: Nitric-Oxide Synthase – Messenger RNA Expression of Different Isoforms in Human Monocytes/Macrophages. Eur. J. Immunol. **24**, 1941 (1994).

626. WEINBERG, J.B., M.A. MISUKONIS, P.J. SHAMI, S.N. MASON, D.L. SAULS, W.A. DITTMAN, E.R. WOOD, G.K. SMITH, B. MCDONALD, K.E. BACHUS, A.F. HANEY, and D.L. GRANGER: Human Mononuclear Phagocyte Inducible Nitric Oxide Synthase (iNOS) – Analysis of iNOS Messenger RNA, iNOS Protein, Biopterin, and Nitric Oxide Production by Blood Monocytes and Peritoneal Macrophages. Blood **86**, 1184 (1995).

627. EVANS, T., A. CARPENTER, H. KINDERMAN, and J. COHEN: Evidence of Increased Nitric Oxide Production in Patients with the Sepsis Syndrome. Circ. Shock **41**, 77 (1993).

628. GOMEZ-JIMENEZ, J., A. SALGADO, M. MOURELLE, M.C. MARTIN, R.M. SEGURA, R. PERACAULA, and S. MONCADA: L-Arginine – Nitric Oxide Pathway in Endotoxemia and Human Septic Shock. Crit. Care Med. **23**, 253 (1995).

629. LORENTE, J.A., L. LANDIN, R. DE PABLO, E. RENES, and D. LISTE: L-Arginine Pathway in the Sepsis Syndrome. Crit. Care Med. **21**, 1287 (1993).

630. WRIGHT, C.E., D.D. REES, and S. MONCADA: Protective and Pathological Roles of Nitric Oxide in Endotoxin Shock. Cardiovasc. Res. **26**, 48 (1992).

631. SZABO, C.: Alterations in Nitric Oxide Production in Various Forms of Circulatory Shock. New Horizons **3**, 2 (1995).

632. SZABO, C., J.A. MITCHELL, C. THIEMERMANN, and J.R. VANE: Nitric Oxide-Mediated Hyporeactivity to Noradrenaline Precedes the Induction of Nitric Oxide Synthase in Endotoxin Shock. Brit. J. Pharmacol. **108**, 786 (1993).

633. THIEMERMANN, C.: Inhibition of Nitric Oxide Synthase Activity in Circulatory Shock: Friend or Foe? In: Role of Nitric Oxide in Sepsis and ARDS (M.P. FINK, and D. Payer, eds.), pp. 201-206. Berlin: Springer. 1995.

634. WEI, X., I.G. CHARLES, A. SMITH, J. URE, G. FENG, F. HUANG, D. XU, W. MULLER, S. MONCADA, and F.Y. LIEW: Altered Immune Responses in Mice Lacking Inducible Nitric Oxide Synthase. Nature, **375**, 408 (1995).

635. LAUBACH, V.E., E.G. SHESELY, O. SMITHIES, and P.A. SHERMAN: Mice Lacking Inducible Nitric Oxide Synthase Are Not Resistant to Lipopolysaccharide Induced Death. Proc. Natl. Acad. Sci. USA, **92**, 10688 (1995).

636. MACMICKING, J.D., C. NATHAN, G. HOM, N. CHARTRAIN, D.S. FLETCHER, M. TRUMBAUER, K. STEVENS, Q.W. XIE, K. SOKOL, N. HUTCHINSON, H. CHEN, and J.S. MUDGETT: Altered Responses to Bacterial Infection and Endotoxic Shock in Mice Lacking Inducible Nitric Oxide Synthase. Cell **81**, 641 (1995).

637. PAYEN, D., C. BERNARD, and S. BELOUCIF: Nitric Oxide in Sepsis. Clinics In Chest Med. **17**, 333 (1996).

638. ALBINA, J.E., C.D. MILLS, W.L. HENRY, and M.D. CALDWELL: Temporal Expression of Different Pathways of Arginine Metabolism in Healing Wounds. J. Immunol. **144**, 3877 (1990).

639. PAYA, D., G.A. GRAY, I. FLEMING, and J.C. STOCLET: Effect of Dexamethasone on the Onset and Persistence of Vascular Hyporeactivity Induced by Escherichia Coli Lipopolysaccharide in Rats. Circ. Shock **41**, 103 (1993).

640. BONE, R.C., C.J. FISHER, T.P. CLEMMER, G.J. SLOTMAN, C.A. METZ, and R.A. BALK: A Controlled Clinical Trial of High Dose Methylprednisolone in the Treatment of Severe Sepsis and Septic Shock. New Eng. J. Med. **317**, 653 (1987).

641. HINSHAW, L., P. PEDUZZI, E. YOUNG, C. SPRUNG, C. SHATNEY, J. SHEAGREN, M. WILSON, and C. HAAKENSON: Effect of High-Dose Glucocorticoid Therapy on Mortality in Patients with Clinical Signs of Systemic Sepsis. New England J. Med. **317**, 659 (1987).

642. HOLLENBERG, S.M., R.E. CUNNION, and J. ZIMMERBERG: Nitric Oxide Synthase Inhibition Reverses Arteriolar Hyporesponsiveness to Catecholamines in Septic Rats. Am. J. Physiol. **264**, H-660 (1993).

643. TEALE, D.M., and A.M. ALLEISTER: Inhibition of Nitric Oxide Synthase Improves Survival in a Murine Persistis Model of Sepsis that is not Cured by Antibiotics Alone. J. Antimicrob. Chemother. **30**, 839 (1992).

644. COBB, J.P., C. NATANSON, W.D. HOFFMAN, R.F. LODATO, S. BANKS, C.A. KOEV, M.A. SOLOMON, R.J. ELIN, J.M. HOSSEINI, and R.L. DANNER: N-Amino-L-Arginine, an Inhibitor of Nitric Oxide Synthase, Raises Vascular Resistance but Increases Mortality Rates in Awake Canines Challenged with Endotoxin. J. Exp. Med. **176**, 1175 (1992).

645. ROBERTSON, F.M., P.J. OFFNER, D.P. CICERI, W.K. BECKER, B.A. PRUITT, and T.R. BILLIAR: Detrimental Hemodynamic Effects of Nitric Oxide Synthase Inhibition in Septic Shock. Arch. Surg. **129**, 149 (1994).

646. PETROS, A., D. BENNETT, and P. VALLANCE: Effects of Nitric Oxide Synthase Inhibitors on Hypotension in Patients with Septic Shock. Lancet **338**, 1557 (1991).

647. PREISER, J.C., P. LEJEUNE, A. ROMAN, E. CARLIER, D. DE BACKER, M. LEEMAN, R.J. KAHN, and J.L. VINCENT: Methylene Blue Administration in Septic Shock – A Clinical Trial. Crit. Care Med. **23**, 259 (1995).

648. LORENTE, J.A., L. LANDIN, R. DE PABLO, E. RENES, and D. LISTE: L-Arginine Pathway in the Sepsis Syndrome. Crit. Care Med. **21**, 1287 (1993).

649. LIU, S.F., I.M. ANDRACLE, R.W. OLD, P.J. BARNES, and T.W. EVANS: Differential Regulation of the Constitutive and Inducible Nitric Oxide Synthase mRNA by Lipopolysaccharide Treatment *In Vivo* in the Rat. Crit. Care Med. **24**, 1219 (1996).

650. BRUNTON, T.L.: On the Use of Nitrite of Amyl in Angina Pectoris. Lancet **2**, 97 (1867).

651. MURRELL, W.: Nitroglycerin as a Cure for Angina Pectoris. Lancet **1**, 80 (1879).

652. DE RUBERTIS, F.R., and P.A. CRAVEN: Calcium Independent Modulation of Cyclic GMP and Activities of Guanylate Cyclase by Nitrosamines. Science **193**, 897 (1976).

653. CHUNG, S.J., and H.L. FUNG: Relationship Between Nitroglycerin Induced Vascular Relaxation and Nitric Oxide Production – Probes with Inhibitors and Tolerance Development. Biochem. Pharmacol. **45**, 157 (1993).

654. MULSCH, A., P. MORDVINTCEV, E. BASSENGE, F. JUNG, B. CLEMENT, and R. BUSSE: *In Vivo* Spin Trapping of Glyceryl Trinitrate Derived Nitric Oxide in Rabbit Blood Vessels and Organs. Circulation **92**, 1876 (1995).

655. MUNZEL, T., H.SAYEGH, and D.G. HARRISON: Nitroglycerin Tolerance is Associated with Increases in Vascular Constriction Mediated by Protein-Kinase-C. Circulation **90** (Suppl. I), I-321 (1994).

656. STAMLER, J.S., D.I. SIMON, J.A. OSBORNE, M.E. MULLINS, O. JARAKI, T. MICHEL, D.J. SINGEL, and J. LOSCALZO: S-Nitrosylation of Proteins with Nitric Oxide: Synthesis and Characterisation of Biologically Active Compounds. Proc. Natl. Acad. Sci. USA **89**, 444 (1992).

657. OYAMA, Y., H. KAWASAKI, Y. HATTORI, and M. KANNO: Attenuation of Endothelium-Dependent Relaxation in Aorta from Diabetic Rats. Eur. J. Pharmacol. **132**, 75 (1986).

658. WANG, S.P., M.W. WEST, L.S. DRESNER, J.F. FLEISHHACKER, D.A. DISTANT, C.M. MUELLER, and R.B. WAIT: Effects of Diabetes and Uremia on Mesenteric Vascular Reactivity. Surgery **120**, 328 (1996).

659. FOZARD, J.R., and M.L. PART: Hemodynamic Responses to N-Monomethyl-L-Arginine in Spontaneously Hypertensive and Normotensive Wistar-Kyoto Rats. Brit. J. Pharmacol. **102**, 823 (1991).

660. PANZA, J.A., P.R. CASINO, C.M. KILCOYNE, and A.A. QUYYUMI: Role of Endothelium-Derived Nitric Oxide in the Abnormal Endothelium-Dependent Vascular Relaxation of Patients with Essential Hypertension. Circ. **87**, 1468 (1993).

661. GIL-LONGO, J., D. FDEZGRANDAL, M. ALVAREZ, M. SIEIRA, and F. ORALLO: Study of *In Vivo* and *In Vitro* Resting Vasodilator Nitric Oxide Tone in Normotensive and Genetically Hypertensive Rats. Eur. J. Pharmacol. **310**, 175 (1996).

662. DEHN, G., M.A. THIEL, F.R. BUHLER, and T.F. LUSCHER: Activation of Endothelial L-Arginine Pathway in Resistance Arteries. Effect of Age and Hypertension. Hypertension **16**, 170 (1990).

663. RANDALL, M.D., G.R. THOMAS, and C.R. HILEY: Effect of Destruction of the Vascular Endothelium upon Pressure Flow Relations and Endothelium Dependent Vasodilatation in Resistance Beds of Spontaneously Hypertensive Rats. Clin. Sci. **80**, 463 (1991).

664. SCHLEIFTER, R., F. PERNOT, A. van OVERLOOP, and A. GAIRARD: *In Vivo* Involvement of Endothelium-Derived Nitric Oxide in SHR – Effects of N^G-Nitro-L-arginine Methyl Ester. J. Hypertension Supp. **9**, 5192 (1991).

665. LACOLLEY, P.J., S.J. LEWIS, and M.J. BRODY: L-N-Nitroarginine Produces an Exaggerated Hypertension in Anesthetized SHR. Eur. J. Pharmacol. **197**, 239 (1991).

666. CHOBANIAN, A.V., A.H. LICHTENSTEIN, V. NIDAKHE, C.C. HAUDENSCHULD, R. DRAGO, and C. NICKERSON: Influence of Hypertension on Aortic Atherosclerosis in the Watanabe Rabbit. Hypertension **14**, 203 (1989).

667. SCHWARTZ, S.M., D. deBLOIS, and E.R.M. O'BRIEN: The Intima – Soil for Atherosclerosis and Restenosis. Circ. Res. **77**, 445 (1995).

668. BOSMANS, J.M., H. BULT, C.J. VRINTS, M.M. KOCKX, and A.G. HERMAN: Balloon Angioplasty and Induction of Non-Endothelial Nitric Oxide Synthase in Rabbit Carotid Arteries. Eur. J. Pharmacol. **310**, 163 (1996).

669. JEONG, M.H., W.G. OWEN, J. GREGOIRE, M.E. STAAB, S.S. SRIVATSA, M.L. STEWART, D.R. HOLMES, and R.S. SCHWARTZ: Local Nitric Oxide Donor Delivery Limits Platelet Deposition and Acute Arterial Occlusion. Circulation 94, I-219 (1996).

670. MARKS, D.S., J.A. VITA, J.D. FOLTS, J.F. KEANEY, G.N. WELCH, and J. LOSCALZO: Inhibition of Neointimimal Poliferation in Rabbits after Vascular Injury by a Single Treatment with a Protein Adduct of Nitric Oxide. J. Clin. Invest. **96**, 2630 (1995).

671. KOIDE, Y., J.D. LUCCA, and A.G. SCICLI: *N*-Acetylcysteine Decreases Neointima Formation After Balloon Injury of the Rat Carotid Artery. Circulation **94**, I-403 (1996).

672. CONRAD, K.P., and K.A. VERNIER: Plasma Level, Urinary Excretion, and Metabolic Production of cGMP During Gestation in Rats. Am. J. Physiol. **257**, R847 (1989).

673. CONRAD, K.P., G.M. JOFFE, H. KRUSZYNA, R. KRUSZYNA, L.G. ROCHELLE, R.P. SMITH, J.E. CHAVEZ, and M.D. MOSHER: Identification of Increased Nitric Oxide Biosynthesis during Pregnancy in Rats. FASEB **7**, 566 (1993).

674. YALLAMPALLI, C., H. IZUMI, M. BYAMSMITH, and R.E. GARFIELD: An L-Arginine-Nitric Oxide-Cyclic Guanosine Monophosphate System Exists in the Uterus and Inhibits Contractility During Pregnancy. Am. J. Obstet. Gynecol. **170**, 175 (1994).

675. YALLAMPALLI, C., R.E. GARFIELD, and M. BYAMSMITH: Nitric Oxide Inhibits Uterine Contractility during Pregnancy but not during Delivery. Endocrinology **133**, 1899 (1993).

676. SOORANNA, S.R., D. BURSTON, B. RAMSAY, and P.J. STEER: Free Amino Acid Concentrations in Human First and 3rd Trimester Placental Villi. Placenta **15**, 747 (1994).

677. BUHIMSCHI, I., C. YALLAMPALLI, Y.L. DONG, and R.E. GARFIELD: Involvement of a Nitric Oxide-Cyclic Guanosine- Monophosphate Pathway in Control of Human Uterine Contractility during Pregnancy. Am. J. Obstet. Gynecol. **172**, 1577 (1995).

678. HULL, A.D., C.R. WHITE, and W.J. PEARCE: Endothelium-Derived Relaxing Factor and Cyclic GMP-Dependent Vasorelaxation in Human Chorionic Plate Arteries. Placenta **15**, 365 (1994).

679. CHANG, J.K., C. ROMAN, and M.A. HEYMANN: Effect of Endothelium-Derived Relaxing Factor Inhibition on the Umbilical-Placental Circulation in Fetal Lambs Inutero. Am. J. Obstet. Gynecol. **166**, 727 (1992).

680. MYATT, L., A.S. BREWER, G. LANGDON, and D.E. BROCKMAN: Attenuation of the Vasoconstrictor Effects of Thromboxane and Endothelin by Nitric Oxide in the Human Fetal-Placental Circulation. Am. J. Obstet. Gynecol. **166**, 224 (1992).

681. SORENSEN, T.K., T.R. EASTERLING, K.L. CARLSON, D.A. BRATENG, and T.J. BENEDETTI: The Maternal Hemodynamic Effect of Indomethacin in Normal Pregnancy. Obstet. Gynecol. **79**, 661 (1992).

682. ROBERTS, J.M., and C.W.G. REDMAN: Preeclampsia – More Than Pregnancy-Induced Hypertension. Lancet **341**, 1447 (1993).

683. SELIGMAN, S.P., J.P. BUYON, R.M. CLANCY, B.K. YOUNG, and S.B. ABRAMSON: The Role of Nitric Oxide in the Pathogenesis of Pre-Eclampsia. Am. J. Obstet. Gynecol. **171**, 944 (1994).

684. BAKER, A.B.: Management of Severe Pregnancy Induced Hypertension, or Gestosis, with Sodium Nitroprusside. Anaesth. Int. Care **18**, 361 (1990).

685. WASSERSTRUM, N.: Nitroprusside in Pre-Eclampsia – Circulatory Distress and Paradoxical Bradycardia. Hypertension **18**, 79 (1991).

686. FARACI, F.M., and J.E. BRIAN: Nitric Oxide and the Cerebral Circulation. Stroke **25**, 692 (1994).

687. IADECOLA, C.: Regulation of the Cerebral Microcirculation during Neural Activity – Is Nitric-Oxide the Missing Link? Trends Neurosci. **16**, 206 (1993).

688. DAWSON, V.L., T.M. DAWSON, E.D. LONDON, D.S. BREDT, and S.H. SNYDER: Nitric Oxide Mediates Glutamate Neurotoxicity in Primary Cortical Cultures. Proc. Natl. Acad. Sci. USA, **88**, 6368 (1991).

689. Sato, S., T. Tominaga, T. Ohnishi, and S.T. Ohnishi: Electron Paramagnetic Resonance Study on Nitric Oxide Production during Brain Focal Ischemia and Reperfusion in the Rat. Brain Res. **647**, 91 (1994).

690. Zheng, G.Z., M. Chopp, F. Bailey and T. Malinski: Nitric Oxide Changes in the Rat Brain after Transient Middle Cerebral Artery Occlusion. J. Neurol. Sci. **128**, 22 (1995).

691. Hamada, Y., T. Hayakawa, H. Hattori, and H. Mikawa: Inhibitor of Nitric Oxide Synthesis Reduces Hypoxic Ischemic Brain Damage in the Neonatal Rat. Pediatr. Res. **35**, 10 (1994).

692. Nishikawa, T., J.R. Kirsch, R.C. Koehler, D.S. Bredt, S.H. Snyder, and R.J. Traystman: Effect of Nitric Oxide Synthase Inhibition on Cerebral Blood Flow and Injury Volume During Focal Ischemia in Cats. Stroke **24**, 1717 (1993).

693. Shapira, S., T. Kedar, and B.A. Weissmann: Dose-Dependent Effect of Nitric Oxide Synthase Inhibition Following Transient Forebrain Ischaemia in Gerbils. Brain Res. **668**, 80 (1994).

694. Yamamoto, S., E.V. Golanov, S.B. Berger, and D.J. Reis: Inhibition of Nitric Oxide Synthesis Increases Focal Ischemic Infarction in Rat. J. Cereb. Blood Flow Metab. **12**, 717 (1992).

695. Zhang, F. and C. Iadecda: Nitroprusside Improves Blood Flow and Reduces Brain Damage after Focal Ischaemia. Neuroreport **4**, 559 (1993).

696. Zhang, F.Y., and C. Iadecola: Reduction of Focal Cerebral Ischemic Damage by Delayed Treatment with Nitric Oxide Donors. J. Cereb. Blood Flow Metab. **14**, 574 (1994).

697. Dalkara, T., T. Yoshida, K. Irikura, and M.A. Moskowitz: Dual Role of Nitric Oxide in Focal Cerebral Ischemia. Neuropharmacol. **33**, 1447 (1994).

698. Yoshida, T., V. Limmroth, K. Irikura, and M.A. Moskowitz: The NOS Inhibitor, 7-Nitroindazole, Decreases Focal Infarct Volume but not the Response to Topical Acetylcholine in Pial Vessels. J. Cereb. Blood Flow Metab. **14**, 924 (1994).

699. Ralston, S.H., L.P. Ho, M.H. Helfrich, P.S. Grabowski, P.W. Johnston, and N. Benjamin: Nitric Oxide – A Cytokine-Induced Regulator of Bone Resorption. J. Bone Miner. Res. **10**, 1040 (1995).

700. McCartney-Francis, N., J.B. Allen, D.E. Mizel, J.E. Albina, Q.W. Xie, C.F. Nathan, and S.M. Wahl: Suppression of Arthritis by an Inhibitor of Nitric Oxide Synthase. J. Exp. Med. **178**, 749 (1993).

701. Ialenti, A., S. Moncada, and M. Dirosa: Modulation of Adjuvant Arthritis by Endogenous Nitric Oxide. Brit. J. Pharmacol. **110**, 701 (1993).

702. Grabowski, P.S. and S.H. Ralston: Cellular Sources of Nitric Oxide Production in the Joint. Br. J. Rheumatol. **34**, 51 (1995).

703. Sakurai, H., H. Kohsaka, M.F. Liu, H. Higashiyama, Y. Hirata, K. Kanno, I. Saito, and N. Miyasaka: Nitric Oxide Production and Inducible Nitric Oxide Synthase Expression in Inflammatory Arthritis. J. Clin. Invest. **96**, 2357 (1995).

704. Farrell, A.J., D.R. Blake, R.M.J. Palmer, and S. Moncada: Increased Concentrations of Nitrite in Synovial Fluid and Serum Samples Suggest Increased Nitric Oxide Synthesis in Rheumatic Diseases. Ann. Rheum. Dis. **51**, 1219 (1992).

705. Grabowski, P.S., A.J. England, R. Dykhuizen, M. Copland, N. Benjamin, D.M. Reid, and S.H. Ralston: Elevated Nitric Oxide Production in Rheumatoid Arthritis – Detection Using the Fasting Urinary Nitrate / Creatinine Ratio. Arth. Rheum. **39**, 643 (1996).

706. Ueki, Y., S. Miyake, Y. Tominaga, and K. Eguchi: Increased Nitric Oxide Levels in Patients with Rheumatoid Arthritis. J. Rheumatol. **23**, 230 (1996).

707. ST. CLAIR, E.W., W.E. WILKINSON, T. LANG, L. SANDERS, M.A. MISUKONIS, G.S. GILKESON, D.S. PISETSKY, D.L. GRANGER, and J.B. WEINBERG: Increased Expression of Blood Mononuclear Cell Nitric Oxide Synthase Type-2 in Rheumatoid Arthritis Patients. J. Exp. Med. **184**, 1173 (1996).

708. ABU-SOUD, H.M., A. PRESTA, B. MAYER, and D.J. STUEHR: Analysis of Neuronal NO Synthase Under Single-Turnover Conditions: Conversion of N^ω-Hydroxyarginine to Nitric Oxide and Citrulline. Biochemistry **36**, 10811 (1997).

709. ABU-SOUD, H.M., R. GACHHUI, F.M. RAUSHEL, and D.J. STUEHR: The Ferrous-dioxy Complex of Neuronal Nitric Oxide Synthase: Divergent Effect of L-Arginine and Tetrahydrobiopterin on Its Stability. J. Biol. Chem. **272**, 17349 (1997).

710. MAYER, B., C.Q. WU, A.C.F. GORREN, S. PFEIFFER, K. SCHMIDT, P. CLARK, D.J. STUEHR, and E.R. WERNER: Tetrahydrobiopterin Binding to Macrophage Inducible Nitric Oxide Synthase: Heme Spin Shift and Dimer Stabilization by the Potent Pterin Antagonist 4-Amino-Tetrahydrobiopterin. Biochemistry **36**, 8422 (1997).

711. GHOSH, D.K., C.Q. WU, E. PITTERS, M. MOLONEY, E.R. WERNER, B. MAYER, and D.J. STUEHR: Characterization of the Inducible Nitric Oxide Synthase Oxygenase Domain Identifies a 49 Amino Acid Segment Required for Subunit Dimerization and Tetra-hydrobiopterin Interaction. Biochemistry **36**, 10609 (1997).

712. STEVENS-TRUSS, R., K. BECKINGHAM, and M.A. MARLETTA: Calcium Binding Sites of Calmodulin and Electron Transfer by Neuronal Nitric Oxide Synthase. Biochemistry **36**, 12337 (1997).

713. JU, H., R. ZOU, V.J. VENEMA, and R.C. VENEMA: Direct Interaction of Endothelial Nitric Oxide Synthase and Caveolin-1 Inhibits Synthase Activity. J. Biol. Chem. **272**, 18522 (1997).

714. GARCÍA-CARDENA, G., P. MARTASEK, B.S.S. MASTERS, P.M. SKIDD, J. COUET, S.W. LI, M.P. LISANTI, and W.C. SESSA: Dissecting the Interaction Between Nitric Oxide Synthase (NOS) and Caveolin: Functional Significance of the NOS Caveolin Binding Domain *in vivo*. J. Biol. Chem. **272**, 25437 (1997).

715. MICHEL, J.B., O. FERON, K. SASE, P. PRABHAKAR, and T. MICHEL: Caveolin *versus* Calmodulin: Counterbalancing Allosteric Modulators of Endothelial Nitric Oxide Synthase. J. Biol. Chem. **272**, 25907 (1997).

716. VENEMA, V.J., R. ZOU, H. JU, M.B. MARRERO, and R.C. VENEMA: Caveolin-1 Detergent Solubility and Association with Endothelial Nitric Oxide Synthase is Modulated by Tyrosine Phosphorylation. Biochem. Biophys. Res. Commun. **236**, 155 (1997).

717. CRANE, B.R., A.S. ARVAL, R. GACHHUI, C.Q. WU, D.K. GHOSH, E.D. GETZOFF, D.J. STUEHR, and J.A. TAINER: The Structure of Nitric Oxide Synthase Oxygenase Domain and Inhibitor Complexes. Science **278**, 425 (1997).

718. SHEARER, B.G., S.L. LEE, K.W. FRANZMANN, H.A.R. WHITE D.C.J. SANDERS, R.J. KIFF, E.P. GARVEY, and E.S. FURFINE: Conformationally Restricted Arginine Analogues as Inhibitors of Human Nitric Oxide Synthase. Bioorg. Med. Chem. Lett. **7**, 1763 (1997).

719. DANDLIKER, P.J., F. DIEDERICH, A. ZINGG, J.P. GISSELBRECHT, M. GROSS, A. LOUATI, and E. SANFORD: Dendrimers with Prophyrin Cores: Synthetic Models for Globular Heme Proteins. Helv. Chim. Acta **80**, 1773 (1997).

720. TAYLOR, P. J. and A. R. WAIT: Sigma-1 Values for Heterocycles. J. Chem. Soc., Perkin Trans II, 1765 (1986).

721. PORTER, T. D. and M. J. COON: Cyctochrome P-450 – Multiplicity of Isoforms, Substrates, and Catalytic and Regulatory Mechanisms. J. Biol. Chem. **266**, 13469 (1991).

(Received December 8, 1997)

Author Index

Page numbers printed in *italics* refer to References

Caveolin-1 139
Chapsyn-110 87, 89
Chloroperoxidase 96
Cholesterol 9, 93
Circular dichroism 115
Citrulline 10, 130
L-Citrulline 23–26, 28, 34, 39, 44, 49–
 54, 56, 57, 59–61, 63, 64, 68–71, 73, 79,
 94, 96, 101, 105, 113, 120, 122, 138
Clostridium spirogenes 13
Complementary deoxyribonucleic
 acid 31, 72
Congenital heart diseases 125
Copper 17, 20, 21
Coronary heart disease 133
Cyanide 80, 100, 107, 132
Cyclic guanosine monophosphate 6, 8,
 14–16, 125, 134
Cyclophilins 85
N^G-Cyclopropyl-L-arginine 62, 63
Cysteine 20
Cytochrome P450-mediated hydroxyla-
 tion 36, 37, 39
Cytochrome P450 reductase 29, 31, 37,
 72, 73, 75–77, 80, 106
Cytochromes P450 28–33, 36, 38, 42, 44,
 45, 48, 49, 55, 56, 59, 62, 94–96, 106,
 107, 142
Cytokine activity 136
Cytokines 10, 22, 33
Cytostatic activity 34
Cytotoxic activity 10–14, 24, 34

N-Dealkylation 39–42, 62
Debrisoquine 42
Debrisoquine 4-hydroxylase 42
N-Demethylation 67, 68
L-Deprenyl 42
Dexamethasone 129
Diabetes 134
Diabetes mellitus 136
2,4-Diamino-6-hydroxypyrimidine 117
Diarrhoea 10
Dihydrobiopterin 108
7,8-Dihydrobiopterin 109
Dihydrofolate reductase 111–113
Dihydropteridine reductase 108
Dimethyl-L-arginine 34
3,5-Dimethylpyridine 99
Dioxygen 11

Dioxygenases 23
Dithionite 76, 77
Dithiothreitol 105, 116
Dystroglycans 87
Dystrophin 82, 85, 87, 89

Electron paramagnetic resonance 77, 78,
 90, 95, 96, 100, 102
Endothelial arginine 124
Endothelial cells 11
Endothelial dysfunction 125, 133
Endothelial nitric oxide synthase 33–35,
 75, 78, 81, 82, 90–94, 96, 98–103, 107,
 111, 112, 115–119, 122, 123, 126, 129–
 131, 135, 136, 138, 139
Endothelin-1 130
Endothelium-derived relaxing factor 6, 7,
 14, 16, 18, 19, 27, 35
Endotoxic shock 11
Endotoxin 129, 130
Epoxidation 43
Escherichia coli 115, 117, 119
Estrone 43–45
Ethylenediamine tetraacetic acid 17, 138
4-Ethylpyridine 101

Ferricyanide 76, 77, 98
Fibrosis 130
Flavine adenine dinucleotide 29–33, 72,
 75, 122
Flavine mononucleotide 29–33, 72, 75,
 77, 78, 122, 124
Flavins 25, 26, 30, 32, 46, 66, 80, 82,
 116, 117
Formaldehyde 60, 61, 63, 69
Formate 44

Glucocorticosteroids 131
Glutamate 14, 15, 83, 101
Glutamate toxicity 136
Glutamic acid 75, 112, 113, 140, 141
γ-Glutamyltranspeptidase 20
Glutathione 20, 21
Glyceryl trinitrate 7, 22, 132, 135
Guanidine 36, 49, 100
Guanidoximes 26, 51, 57, 65, 107
Guanosine triphosphate 6, 8
Guanylate cyclase 6–9, 18, 19, 26, 27,
 34, 119
Guanylate kinase 89

SpringerChemistry

Fortschritte der Chemie organischer Naturstoffe

Progress in the Chemistry of Organic Natural Products

Founded by L. Zechmeister
Editors: W. Herz, H. Falk, G. W. Kirby,
             R. E. Moore, and Ch. Tamm

Volume 75

1998. VII, 226 pages. 26 figures.
Hardcover DM 250,–, öS 1750,–, US $ 159.00
Subscription price: Hardcover DM 225,–, öS 1575,–
ISBN 3-211-83053-7

Contents:
Cyclopeptide Alkaloids
(D.C. Gournelis, G.G. Laskaris, and R. Verpoorte)
Naturally Occurring 6-Substituted 5,6-Dihydro-Alpha-Pyrones
(L.A. Collett, M.T. Davies-Coleman, and D.E.A. Rivett)

Volume 74

1998. VII, 300 pages. 5 figures.
Hardcover öS 2030,–, DM 290,–, US $ 199.00
Subscription price: Hardcover öS 1827,–, DM 261,–
ISBN 3-211-83033-2

Contents:
Triterpenoid Saponins
(S. B. Mahato and S. Garai)
Introduction • Isolation • Structure Elucidation • Biological
Activity • Production of Saponins by Tissue Culture• Future
Possibilities • Reports of New Triterpenoid Saponins
Synthesis of 6-Deoxyamino Sugars
(L. A. Otsomaa and A. M. P. Koskinen)
Introduction • Known 6-Deoxyaminohexoses • Synthetic aspects

 SpringerWienNewYork

Sachsenplatz 4-6, P.O.Box 89, A-1201 Wien, Fax +43-1-330 24 26, e-mail: order@springer.at, Internet: http://www.springer.at
New York, NY 10010, 175 Fifth Avenue • D-14197 Berlin, Heidelberger Platz 3 • Tokyo 113, 3-13, Hongo 3-chome, Bunkyo-ku

SpringerChemistry

Amino
Acids

Editors-in-Chief: G. Lubec, Wien
G. C. Barrett, Oxford
R. M. Williams, Fort Collins, CO
and an International Editorial Board

Aim and Scope

Amino Acids publishes contributions from all fields of amino acid research: analysis, separation, synthesis, biosynthesis, cross linking amino acids, racemization/enantiomers, modification of amino acids as phosphorylation, methylation, acetylation, glycosylation and nonenzymatic glycosylation, new roles for amino acids in physiology and pathophysiology, biology, amino acid analogues and derivatives, polyamines, radiated amino acids, peptides, stable isotopes and isotopes of amino acids. Applications in medicine, food chemistry, nutrition, gastroenterology, nephrology, neurochemistry, pharmacology, excitatory amino acids are just some topics to be listed. We also encourage the submission of papers of interdisciplinary borderlines.

Subscription information

1999. Volumes 16+17 (4 issues each):
DM 1168,–, öS 8176,–, plus carriage charges
US $ approx. 754.00 including carriage charges
ISSN 0939-4451, Title No. 726

Subscription price for members of the International Society
for Amino Acids Research: US $ 126.00 + US $ 52.00 carriage charges
(Orders have to be sent directly to Springer-Verlag)

For further information please visit our homepage.
View table of contents and abstracts online at:
http://www.springer.at/amino_acids.

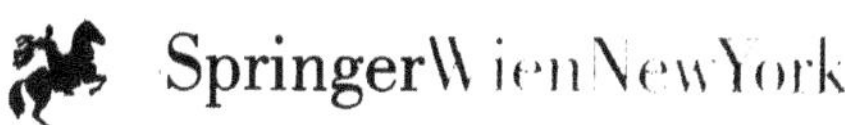

SpringerWienNewYork

Sachsenplatz 4-6, P.O.Box 89, A-1201 Wien, Fax +43-1-330 24 26, e-mail: order@springer.at, Internet: http://www.springer.at
New York, NY 10010, 175 Fifth Avenue • D-14197 Berlin, Heidelberger Platz 3 • Tokyo 113, 3-13, Hongo 3-chome, Bunkyo-ku

GPSR Compliance
The European Union's (EU) General Product Safety Regulation (GPSR) is a set
of rules that requires consumer products to be safe and our obligations to
ensure this.

If you have any concerns about our products, you can contact us on

ProductSafety@springernature.com

In case Publisher is established outside the EU, the EU authorized
representative is:

Springer Nature Customer Service Center GmbH
Europaplatz 3
69115 Heidelberg, Germany